# Negative Ions and
# the Magnetron

# Negative Ions and the Magnetron

F. M. Page

*Professor, Department of Chemistry,*
*University of Aston in Birmingham*

G. C. Goode

*Research Fellow,*
*University of Sheffield*

WILEY–INTERSCIENCE

a division of John Wiley & Sons Ltd.,

London   New York   Sydney   Toronto

539.721
P14n
71242
September, 1970

Made and printed in Great Britain by
William Clowes and Sons, Limited
London and Beccles

# Preface

The study of negative ions, and of their stabilities, has been over-shadowed by the more numerous and easily studied positive ions. Most phenomena which could give information about negative ions are dominated by the lightest negative ion, the electron, and the difficulty of disentangling the effects due to the electron has held back a systematic study of heavy negative ions. This neglect is quite unjustifiable, and the study of negative ions is just as important and rewarding as the study of positive ions. We hope that this monograph will encourage the growth of work in this field.

The general reviews of the stabilities of negative ions due to Pritchard and Buchelnikova, augmented in certain areas by the essays of Branscomb and Brieglebs, collect together most of the few experimental results available until recently, and give an account of the methods used to obtain these results. The present work is restricted to the results obtained by only one method, that due to Mayer, but examines this method in detail.

This detailed examination takes the method far beyond the original application to the determination of electron affinities, and involves the fields of surface chemistry, adsorption of gases and theory of rate processes. The results of this examination have enabled some rules for the semi-empirical prediction of electron affinities to be laid down.

We have collected the experimental results, many of them hitherto unpublished, from our colleagues and present them together in the hope that they will be of service to other workers. The results are listed in detail in the main body of the text, and according to the particular ion, or substance examined, in the appendices. We have used kilocalories throughout the text, but appendix II gives additionally the energies expressed in kilojoules and electron-volts.

We would wish to pay tribute to the generous support received from the Ministry of Technology, the U.S. Army, the Institute of Petroleum, the Royal Society and the Science Research Council; to the many industrial concerns, laboratories or individuals who have so generously given us samples of material; to the Universities of Aston and Cambridge where the work has been carried out, and to our colleagues who are named within this work, but in particular Drs. Farragher, Gaines and Kay, who contributed so much to the final achievement.

G.C.G.
F.M.P.

*Aston,*
*August 1968*

# List of Symbols

| | |
|---|---|
| $A$ | Anode |
| $A$ | Area |
| $A$ | Constant |
| $B$ | Magnetic flux density |
| $B_c$ | Critical value of magnetic flux density |
| $B$ | Constant in Richardson's equation |
| $B$ | Constant |
| $C$ | Constant |
| $C$ | Vibrational heat capacity |
| $C$ | Coulombic energy in charge transfer equation |
| $D_{A-B}$ | Dissociation energy of A—B bond |
| $D$ | Unqualified dissociation energy of A—B bond |
| $E$ | Energy radiated |
| $E_A$ | Electron affinity of A |
| $E_a$ | Activation energy for adsorption |
| $E_d$ | Heat of desorption |
| $E'$ | Apparent electron affinity |
| $E_r, E_\theta$ | Radial and azimuthal electron fields |
| $E_0, E_g$ | Basic and group electron affinities |
| $E_{ix}$ | Apparent electron affinity of group $i$ at distance $x$ |
| $E_i$ | Apparent electron affinity of group $i$ at distance 0 |
| $F$ | Electrostatic force on ion |
| $F$ | Fermi function |
| $G_0, G_i$ | Interaction energies |
| $H$ | Enthalpy |
| $I_x$ | Moment of inertia |
| $I_a$ | Ionization potential of A |
| $I_{ia}$ | Group contribution to $I_a$ at distance $x$ |
| $K$ | Pauling thermochemical factor |
| $K$ | Equilibrium constant |
| $Q^*$ | Reduced partition function |
| $Q_i$ | Partition function |
| $Q_A$ | Heat of adsorption of A |
| $Q_{A/B}$ | Heat of adsorption of A on B |
| $R$ | Gas constant |
| $S_K$ | Kinetic contribution to entropy |

$S_T$     Entropy at temperature T

$T$     Temperature ($T_f$, $T_{fil}$, of filament)

$U$     Electrostatic potential energy

$V$     Potential

$W$     Electrostatic energy

$W$     Internal energy

$W$     Energy change in ion formation

$X_A$     Electronegativity of A

$X_0$     Interaction energy

$a$     Coefficient in C.T. equation

$a$     Radius of cavity in dielectric

$a$     Covalent bond length

$a$     Distance of point of action of dipole from ring

$a_0, a_i$     Coefficient of wave function

$b$     Coefficient in C.T. equation

$b_0, b_i$     Coefficient of wave function

$c_x$     Concentration

$d$     Transmission coefficient

$e$     Electronic charge

$f$     Magnetic force on ion

$h$     Planck's constant

$j_0$     Current from clean surface

$j_e$     Electron current

$j_{fil}$     Filament current

$j_i$     Ion current

$j_t$     Total current

$k$     Constant

$k_1, k_2$     Rate constants in Langmuir's equation

$l$     Length

$l$     Thickness of barrier

$m_e$     Electron mass

$n$     Mullikan electronegativity

$p$     Pressure

$q$     Heat of adsorption

$q_0$     Heat of adsorption at zero coverage

$r$     Distance

$r_a$     Radius of anode

$r_f$     Radius of filament

$r_i$     Distance of centre of $i$th dipole from molecular centre

$r_j$     Distance parameter

$u_a, u_d$     Rates of adsorption and desorption

$v$     Velocity

$x$     Distance

$x$     Reduced vibrational contribution

$x$     Exponent of pressure dependence

$\alpha$     Polarizibility

$\alpha_x$     Polarizibility of $x$

$\alpha_0$     Molecular polarizibility

| | |
|---|---|
| $\alpha$ | Ratio of transmission coefficients |
| $\alpha, \beta$ | Energy levels in ground states |
| $\beta_i$ | Geometrical factor relating $E_i$ and $E_{ix}$ |
| $\Gamma$ | Dielectric constant |
| $\delta$ | Error in temperature measurement |
| $\Delta$ | Pauling thermochemical differences |
| $\Delta$ | Difference symbol |
| $\theta$ | Azimuthal angle |
| $\theta$ | Fraction of surface covered |
| $\theta_i$ | Fractional area of $i$th patch |
| $\mu$ | Dipole moment |
| $\mu_i$ | Dipole moment of $i$th group |
| $\nu$ | Frequency |
| $\nu_{ct}$ | Frequency of C.T. band maximum |
| $\sigma$ | Stefan's constant |
| $\sigma$ | Ground state energy of ion |
| $\Phi_i$ | Fraction of current for $i$th patch |
| $\chi$ | Work function |
| $\chi'$ | Experimental work function |
| $\chi_i$ | Work function of $i$th patch |
| $\bar{\chi}$ | Mean work function |
| $\chi_m$ | Measured thermonic work function |
| $\chi_0$ | Experimental work function of clean surface |
| $\psi_0, \psi_i$ | Wave function |
| $\omega$ | Angular velocity |

# Contents

# The Modes of Formation and Importance of Negative Ions

## 1.1  Introduction

The study of the energetics of charged particles has increased in scope during the past thirty years with the development of instruments, such as the mass spectrometer, which are specifically designed to handle and monitor charged species, and with the realization that a great many processes involving electron transfer may lead to the production of such species. Since charged particles may be moved from place to place electromagnetically, even in the presence of a great excess of neutral species, it is possible to study specific processes involving ions with great ease. Furthermore, many charged species which are intrinsically unstable may be stabilized by interaction either with charges of opposite sign, as in a crystal lattice, or with polarized matter, as in a solution. Charged species are therefore of importance in many fields of science, from astrophysics to biology, and a knowledge of their fundamental stability adds to our understanding of these fields.

## 1.2  Ionization Potential and Electron Affinity

The attachment or removal of an electron from an atom, molecule or radical entails a change in the total energy of the system. If such atom, molecule or radical is a neutral species, then work must be done on the system to remove an electron, and this increase in energy relative to the ground state consequent upon the removal of the electron to an infinite distance from the positively charged residue is known as the *ionization potential*. Similarly if an electron is brought up to the neutral species to form a negative ion the system decreases in energy if the negative ion is stable, and this decrease in energy relative to the infinitely separated molecule and electron is called the *electron affinity*. The two terms are, in the general case, interchangeable, and it would be reasonable to talk of the ionization potential of a negative ion as being identical with the electron affinity of the neutral species, although the electron is being removed and not added. Thermodynamically, the definition of ionization potential, which gives a positive sign to an increase in the energy of a system, is to be preferred to that of electron affinity, but since it is customary to restrict the term ionization potential to processes resulting in the formation of positive ions, the term *stability* of the negative ion will be used. A positive stability then implies that work has to be done on the system to detach the electron. The term stability, being general, could lead to confusion, but the stability of the negative ion with respect to any mode of decomposition other than that of loss of the electron will always be qualified by the specification of the mode of decomposition. The term electron affinity will be retained for discussion of the properties of the neutral species, particularly if the negative ion concerned has not been identified.

## 1.3 Electron Affinities of Molecules

The term molecule is here used in the most general way. It includes not only atoms, radicals and 'closed-shell' molecules, but also positive and negative ions, which may have further charges added to them.

It is conceptually easy to visualize negative ions formed from radicals, where the extra electron satisfies the valence requirements of the accepting atom, but many molecules which have closed shells, e.g. $SF_6$, may still attach an electron to form a negative ion; it does not always follow that the stabilities of these negative ions must be positive although other factors favouring the charged state of the system would have to be important if an ion of negative stability were to be observed. If we examine the potential energy curves of Figure 1.1 we notice that for the hypothetical ions $HI^-$ and $HCl^-$ there must be some slight stability with respect to decomposition to H and the halide ion, even if only the van der Waals energy of H and $X^-$. The minimum in these curves therefore lies two or three kcal mole$^{-1}$ below the zero at infinite separation, which in turn must lie below the corresponding zero for the uncharged HX by the electron affinity of the atom X. Since the ground state of the species HX lies below the zero $(H + X)$ by the dissociation energy $D_{HX}$, we can write

$$E_{HX} = E_X + D_{HX^-} - D_{HX}$$

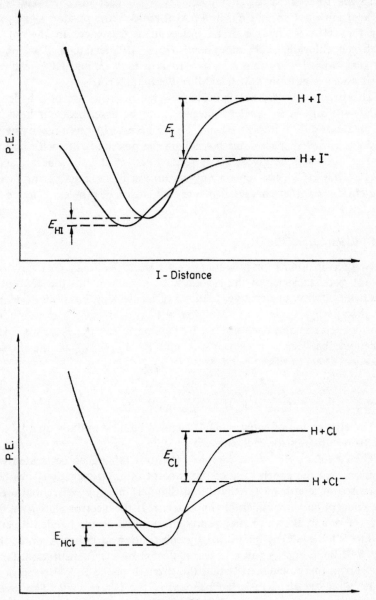

Figure 1.1 Schematic potential energy diagrams for negative molecular ions (above) HI⁻ (below) HCl⁻

In the case of hydrogen iodide, $D_{HI} = 70$ kcal mole$^{-1}$ [1] and $E_I = 71$ kcal mole$^{-1}$ [2] so that $E_{HI}$ is $+5$ kcal mole$^{-1}$, the positive sign implying that HI$^-$ is capable of an independent existence. In the case of hydrogen chloride, on the other hand, $D_{HCl} = 102$ kcal mole$^{-1}$ and $E_{Cl} = 84$ kcal mole$^{-1}$ [3,4] so that $E_{HCl} = -15$ kcal mole$^{-1}$, and HCl$^-$ can only exist under conditions which stabilize the ionic form.

This argument does not indicate how the negative ion of a hydrogen halide (or any other singlet molecule) is to be described, nor how it is formed, nor does it imply that such an ion necessarily has a real existence. It does, however, place some bounds on the possibility of such existence, and in the example given, HI if formed can exist as a separate stable species, but HCl$^-$ can remain only as an ion if some stabilizing field is present due to other charges as in a crystal lattice, or dipoles, as in a polar solvent.

## 1.4 Electronegativity

Neither the ionization potential nor the electron affinity, by themselves, give a good measure of the reactivity of a radical, but the concept of electronegativity, or negative character of an element, has been developed empirically by Pauling[5] and Mulliken[6,7] to provide such a measure. The thermochemical approach utilized by Pauling relates the electronegativity difference between two elements A and B ($X_A - X_B$), to the thermochemical strength of the bond A–B ($D_{A-B}$)

$$X_A - X_B = K\Delta^{\frac{1}{2}} \qquad (1.1)$$

where $\quad \Delta = D_{A-B} - \sqrt{D_{A-A}D_{B-B}}$ in eV $\quad$ and $\quad K = 23^{-\frac{1}{2}}$

The application of equation (1.1) enabled Pauling to draw up a table of electronegativities for each element.

The calculation of bond energies from this table is to be treated with caution, and the general use of the concept of electronegativity has been discussed at length by Pritchard and Skinner[8] but the differential use of the concept has a considerable importance. If the electronegativity difference of A and B (which may represent any two atoms bonded together within a molecule) is unaffected by substitution at either A or B, then the A–B bond energy will also be unaffected by such substitution. Gray[9] has shown thermochemically that the strength of the RO–H bond in the lower aliphatic alcohols is independent of the nature of R. The inverse argument applied to this result indicates that the electronegativity of the oxygen atom is independent of the nature of R.

This observation would be of little quantitative importance in the study of negative ions had not Mulliken[6,10] advanced an alternative formulation for the electronegativity of an atom as the mean of the sum of the ionization potential and electron affinity

$$n_A = \tfrac{1}{2}(I_A + E_A) \tag{1.2}$$

Since many ionization potentials have been measured and tabulated[11], equations (1.1) and (1.2) may be used either to give rough estimates for the electron affinity, if the thermochemistry is understood, or, if the electron affinities are known, to give useful estimates of bond strengths and reactivities of radicals, and this use provides yet another incentive for the study of electron affinities.

## 1.5  Formation of Negative Ions

The formation of a negative ion from a given precursor may proceed through a number of different mechanisms. It may occur homogeneously in the gas phase, or heterogeneously at the surface of a condensed phase, usually at the surface of a heated metal filament.

Table 1.1

| Homogeneous | | |
|---|---|---|
| Thermal dissociation | $AB\,(+X)$ | $\rightarrow A^+ + B^- \,(+X)$ |
| Chemi-ionization | $A + B$ | $\rightarrow A^+ + B^-$ |
| Photoionization | $AB + h\nu$ | $\rightarrow A^+ + B^-$ |
| Charge transfer | $AB + h\nu$ | $\rightarrow A^+ B^-$ |
| Direct capture | $AB + e\,(+X)$ | $\rightarrow AB^- \,(+X)$ |
| Dissociative capture | $AB + e$ | $\rightarrow A + B^-$ |
| Ion pair formation | $AB + e$ | $\rightarrow A^+ + B^- + e$ |
| Charge exchange | $A^- + B$ | $\rightarrow A + B^-$ |

| Heterogeneous | | |
|---|---|---|
| Direct capture | $AB + \text{metal}$ | $\rightarrow AB^-$ |
| Dissociative capture | $AB + \text{metal}$ | $\rightarrow A + B^-$ |
| Double capture | $AB^+ + \text{metal}$ | $\rightarrow AB^-$ |
| Surface ionization | $AB + \text{metal}$ | $\rightarrow A^+ + B^-$ |

In each case of heterogeneous formation, the metal acts as an infinite source of electrons.

All of the processes listed in Table 1.1 can give rise to negative ions, and most of them have been used in studies of the stabilities of negative ions, but few of them are rapid enough to contribute significantly to the formation of negative ions in the magnetron.

### Thermal Dissociation

This process is restricted to the dissociation of salts on energetic grounds, and is unlikely to be a useful source of negative ions, since the neutral dissociated state is of lower energy than the ionic dissociated state (a possible exception would be caesium nitrite). The operation of the non-crossing rule (Figure 1.2) leads to the paradox that the ground state of the sodium chloride molecule is an ion pair, but dissociates to neutral species, while the excited state is covalent in character but dissociates to ions. If an equilibrium can be set up, all the possible components of the sodium chloride system will be present, and an analysis of their proportions can be used to determine the relevant energies[12].

### Chemi-ionization

This process, which being bimolecular would be expected to be fast, has been invoked by Padley, Page and Sugden[13,14] to explain anomalies in the ionization of alkali metals in flames. It is not, however, likely to be a very good source of negative ions because of the large amount of energy needed to produce the positive ion at the same time.

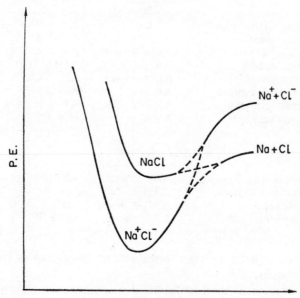

Figure 1.2  Schematic potential energy diagram for the sodium chloride system

## Photoionization

Photoionization, or photochemical ion pair formation, has been shown to occur in the gas phase[15,16] but the energy required to produce the ion pair will be large, just as for thermal dissociation. The heterolytic fission of a covalent bond will involve energies of the order of 7 eV, which necessitates work in the vacuum ultraviolet region. Studies in this region have produced precise information about the stabilities of some negative ions, but the adventitious production of negative ions by this process is not likely to be important in the magnetron, although it may well contribute to the negative ion population of the upper atmosphere.

## Charge Transfer

The basic energy changes contributing to negative ion formation are no smaller than those contributing to photoionization, but since the resulting ion pair remain closely associated, the overall energy change is diminished by the coulombic energy of the ion pair, which may amount to 5 V (110 kcal), so that this photochemical process may be induced by visible light. This process does not contribute to the formation of negative ions in the magnetron, but a great deal of work has been done on the charge transfer spectra of organic complexes in solution[7,17,18,19,20] and scales of electron affinity have been described. In solution, the wavelengths of most intense absorptions are given by

$$h\nu = aI - bE - C \qquad (1.3)$$

where $a$ and $b$ are factors close to unity, and $C$ represents the coulombic energy of the ion pair. A comparison of equation (1.3) with actual values for the electron affinities determined in this work suggests that the coefficient $b$ has a value of approximately $0.72$[21]. The reasons for this are discussed in Chapter 9.

## Electron Induced Processes

The three processes grouped under this heading are those responsible for the production of ions in discharges, or in a mass spectrometer. These processes are of importance at atmospheric pressure, for example in the electron capture detector for gas chromatography[22], and they could be important in the magnetron, where the pressures are not so low that the mean free path of an electron is comparable with the dimensions of the apparatus. Experiments on the movement of electrons through vapours[23,24] indicate that the probability of electron attachment at a collision is about $10^{-4}$, so that it is most improbable that any negative ions will be formed in the vapour in a magnetron.

*Charge Exchange*

Processes involving the transfer of charge between ion and molecule are of considerable importance in producing molecular ions. The rate of a direct capture process will be low, whether it is one of radiative attachment, or of three-body attachment. Both types of process are inefficient, with probabilities of the order of $10^{-4}$, and the three-body process depends to a higher order on the pressure, or number of chaperon molecules present, and is therefore negligible at low pressure in comparison with the dissociative attachment processes.

The capture of an electron at low pressures will therefore lead, preferentially, to fragment ions, even if their stability is low. Molecular ions can, however, be formed from these fragment ions by a process of charge exchange.

$$AB + C^- \rightarrow AB^- + C$$

This process is bimolecular, and efficient, and will therefore be rapid, particularly if the molecular ion $AB^-$ is stable, and the fragment ion $C^-$ is of low stability. The negative charge may be transferred directly, or as part of a more complex chemical rearrangement.

$$NO_2 + O_2^- \rightarrow NO_2^- + O_2 \qquad \text{Electron transfer}$$
$$NO + O_2^- \rightarrow NO_2^- + O \qquad O^- \text{ transfer}$$

Despite the efficiency of the charge exchange processes, they are unlikely to be important in the magnetron, because of absence of collisions in the gas phase. Even if they did occur, they would not upset the conclusions drawn, since the resolution of the magnetron is only sufficient to separate the electrons and heavy ions. The charge exchange process does not alter the total abundance of heavy ions. A similar argument may be advanced to discount the importance, in the magnetron, of processes of ionic rearrangement, or of the breakdown of metastable ions. The abundance of the heavy ions is determined by the process forming the primary ion, and any subsequent processes affecting the nature but not the abundance of the ions are unobservable and therefore irrelevant. The attempts now being made to couple a mass spectrometer to a magnetron will, if successful, alter this by making the identification of each type of ion possible.

*Heterogeneous processes*

These processes are of dominant importance in the experiments to be described, and will be analysed in detail at a later stage. They are fast and efficient, because the hot metal filaments used act as copious sources of electrons, and at the same time are ever-present chaperons, so that stabilization of the ion is achieved.

The only process in this group which is not of great importance is that of double electron capture by positive ions. It is well known that a beam of positive ions incident on a surface can be reflected as a negative beam, but such capture becomes efficient only at positive ion energies of several hundreds of volts, which is well above the energy range of interest in this work.

## 1.6   Negative Ions in Flames and Discharges

One of the fields where a knowledge of the stability of negative ions is most important is that of flame ionization. Flames, and shock-heated gases, are commonly associated, as in rocket propulsion, with electromagnetic attenuation of signals and the control of this attenuation may be effected through a control of the negative ions. An electromagnetic wave, in passing through a partially ionized medium, excites the ions into sympathetic oscillation. The ions rapidly lose the directed momentum thus acquired at subsequent collisions with gas molecules, so that the energy of the wave is degraded into thermal energy of the gas. The only ion which is of significant importance to this process is the electron, by virtue of its low mass, and the process of attenuation may therefore be controlled by converting the free electrons into stable negative ions of much higher mass. The inverse of this problem occurs in magnetohydrodynamic generators, where electricity is generated directly from a high velocity flame between the poles of a magnet. Since the flame is here equivalent to the armature of a dynamo, and the electrical energy of the output is obtained at the expense of the thermal and kinetic energy of the flame gases, it is of vital importance that the conductivity of the flame be high, which implies a high concentration of free electrons. Heavy negative ions are undesirable parasites here, and a knowledge of the stabilities of those negative ions which can be formed from the flame gases or from impurities is required, properly to assess the performance of an M.H.D. generator.

In addition to negative ions of the halogens, or of electronegative impurities, many negative ions have been detected in the natural ionization of undoped hydrocarbon flames[25,26,27]. Some of these ions, such as $OH^-$, $CN^-$, are simple, some could be alkoxide or carboxylate ions, at least on a mass basis, but the most striking feature of the negative ion mass spectrum of a flame is the large number of mass peaks which appear to be due to hydrocarbon negative ions, among which the ions derived from acetylenes are particularly prominent.

Negative ions are formed in flames by attachment of the large number of free electrons produced thermally or chemically in the hot reacting gases. For similar reasons, negative ions are present in electrical discharges, and a suitable discharge may be used, in conjunction with a mass filter, as a source of negative ions in a molecular beam study. Alternatively, since electrons may readily be produced by field ionization, and the very field which produces the electrons will accelerate them to produce a secondary electron avalanche, the presence of an electron acceptor may, by capturing the primary electron, prevent the formation of the avalanche. This is put to good use when electrical switchgear is filled by an electronegative gas such as sulphur hexafluoride. Stable negative ions ($SF_6^-$) are formed, and these are so heavy that all secondary ionization is suppressed, particularly when the exciting field is transient as in the opening of a heavy duty circuit breaker.

## Negative Ions in Biology

A considerable amount of work has been done by Lovelock[22, 28] and his colleagues on the formation of negative ions by biologically active materials. Using the electron capture detector developed by this school, in conjunction with a gas chromatograph, the relative reduction in the electron concentration of a radioactively generated plasma by the various materials was determined. For a constant amount of added material, this is the measurement of sensitivity of the instrument, considered as a gas chromatograph detector. The general trends in the sensitivity to different materials were as expected, bearing in mind the lack of discrimination between ions formed by direct capture and by dissociative capture, and followed the qualitative estimates of the electronegativity of the material (certain groupings enhanced the sensitivity; for example, polysulphide chains and halogen atoms). There were, however, a number of anomalously high sensitivities, and these were almost always associated with a high biological activity. Small changes in structure between isomers showed a close parallel between detector response and biological activity while many biological metabolic intermediates also showed a high detector response. It is not to be asserted that the ability to capture electrons, as measured by the detector response, is a direct cause of biological activity, but the parallel between response and activity indicates that there is a close link between the two.

## Negative Ions in Atmospheric Chemistry and Astrophysics

The facile production of electrons by thermal or photoionization in stars, or in the upper levels of the atmosphere makes the formation of negative ions very probable. The demonstration of the existence of the ionosphere in the early days of wireless telegraphy was soon followed by a study of its diurnal variation, and the deduction that the free electrons produced by sunlight were captured to form negative ions during the night.

Careful observation of the time of the 'radio dawn'[29] shows that it corresponds to the arrival of light of about 3000 Å wavelength, which indicates a detachment energy of about 4 V. Rocket-borne mass spectrometers indicate that the important negative ions at a height of 110 km are, in decreasing order of importance, $NO_2^-$, $NO_3^-$ and $O_2^-$, with only traces of others[30]. Experiments in the laboratory confirm that $NO_2$ has an electron affinity of 4·0 V[31], agreeing with the field observations.

Direct observation of negative ions in stellar atmospheres is not possible, but their existence has been inferred from the spectral energy distribution. The apparent colour temperature of the sun may be accounted for by the presence of significant quantities of $H^-$, while evidence has been brought forward suggesting that $Cl^-$, and even in cooler stars $CN^-$ and $C_2^-$, may also be important.

*Negative Ions in Chemistry*

The importance of an understanding of the stability of negative ions in chemistry may be considered under three headings. It is not the present purpose to consider the chemistry of negative ions as such, even though so much work has been done on them, but rather to outline those fields where the character of the negative ion is subordinate to the field itself. These principal fields are reaction mechanisms, electrode processes and solvation. The majority of chemical reactions which do not take place in a highly polar solvent do not involve charged species, because the energetics forbid such ionization, but in a number of cases it is believed that an ionic mechanism does occur. Rather more are believed to proceed through a positively charged species than through a negative ion, but, in large measure, the lack of information about negatively charged species has concentrated attention upon the positive ions. The part played by the proton in acid catalysed reactions, which usually involve a positively charged species as intermediate, may be played by the free electron in radiation induced reactions, but the solvation of the proton followed by its transfer to a reactant is usually an energetically more favourable step than the solvation of a free electron followed by transfer, and the solvation of the electron usually results in breakdown of the solvent. One group of reactions which do, however, appear to involve the intermediate formation of a negative ion, are the surface catalysed rearrangements of hydrocarbons, and indeed the relative ease of transfer of an electron to and from a surface favours ionic reactions generally.

The extreme form of such ionic interaction with the surface leads to electrochemical reactions, and a great deal of work has been done on such processes, both at ordinary electrodes and at highly polarized electrodes such as the polarograph[32,33,34,35,36,37]. Numerous chemical

changes can be induced by electrochemical reaction, and since the intermediate products, which after a one-electron addition must be free radicals, may be studied conveniently by e.s.r., there is a growing body of
evidence about the structure of these intermediate negative ions and about
the mechanism of the reactions. The quantitative study of the reduction
potentials of neutral species in the polarograph can give precise values
for the energetics of the overall reaction, but since these reactions occur in
solution the observed energetics include terms representing the solvation
energies of the various species. Despite much work on the theories of
solvation, there is little reliable experimental information about the solvation energies of negative ions, so that a direct comparison of the heats
of gaseous ionic processes, based upon measurements of electron affinities
in the gas phase, with the energetics of the same reaction occurring in
solution will yield information about the heats of solvation of ions which
will, in turn, facilitate an understanding of electrode processes.

## REFERENCES

1. Cottrell, T. L., *The Strengths of Chemical Bonds*, Butterworths, London,
   1958.
2. Berry, R. S., Reimann, C. W. and Spokes, G. N., *J. Chem. Phys.*, **37**,
   2278 (1962).
3. Berry, R. S., Reimann, C. W. and Spokes, G. N., *J. Chem. Phys.*, **35**,
   2237 (1961).
4. McCallum, K. J. and Mayer, J. E., *J. Chem. Phys.*, **11**, 56 (1943).
5. Pauling, L., *The Nature of the Chemical Bond and the Structure of Molecules and Crystals*, Cornell University Press, 1960.
6. Mulliken, R. S., *J. Chem. Phys.*, **2**, 782 (1934).
7. Mulliken, R. S. and Person, W. B., *Ann. Rev. Phys. Chem.*, **13**, 107
   (1962).
8. Pritchard, H. O. and Skinner, H. A., *Trans. Faraday Soc.*, **49**, 1254
   (1953).
9. Gray, P., *Trans. Faraday Soc.*, **52**, 344 (1956).
10. Mulliken, R. S., *J. Chem. Phys.*, **3**, 573 (1935).
11. Frankevich, Ye L., Gurvich, L. V., Kondrat'yev, V. N., Medvedev,
    V. A. and Vedeneyev, V. I., *Bond Energies, Ionisation Potentials and
    Electron Affinities*, Arnold, 1966.
12. Tandon, A. N., *Proc. Natl. Acad. Sci., India*, **7**, 102 (1937).
13. Padley, P., Page, F. M. and Sugden, T. M., *Trans. Faraday Soc.*, **57**, 1552
    (1961).
14. Page, F. M. and Woolley, D. E., *Anal. Chem.*, **40**, 210 (1968).
15. Chupka, W. A., *Private communication*.
16. Dibeler, V., *Private communication*.
17. Batley, M. and Lyons, L. E., *Nature*, **196**, 573 (1962).
18. Briegleb, G. and Czekalla, J., *Z. Electrochem.*, **63**, 6 (1959).
19. Briegleb, G., *Angew. Chem.* (Intern. ed.), **3**, 617 (1964).
20. McConnell, H., Ham, J. J. and Platt, J. R., *J. Chem. Phys.*, **21**, 66 (1953).

21. Farragher, A. L., *Ph.D. Thesis*, The University of Aston in Birmingham, 1966.
22. Lovelock, J. E., *Anal. Chem.*, **33**, 162 (1961): Wentworth, W. E. and Chen, E., *J. Phys. Chem.*, **67**, 2201 (1963).
23. Brown, S. C., *Basic Data of Plasma Physics*, The Technology Press of the Massachusetts Institute of Technology, and Wiley, New York, 1961.
24. Massey, H. S. W. and Burhop, E. H. S., *Electronic and Ionic Impact Phenomena*, Oxford University Press, 1952.
25. Feugier, A. and von Tiggelen, A., *Tenth Symp. (Intern.) Combustion*, The Combustion Institute, Pittsburgh, 1965.
26. Green, J. A., *AGARD Conf. Proc. C.P.8.* Vol. 1, pp. 191–214, Sept. 1965.
27. Miller, W. J. and Calcote, H. P., *J. Chem. Phys.*, **41**, 4001 (1964).
28. Lovelock, J. E., Zlatkis, A. and Becker, R. S., *Nature*, **193**, 540 (1962)
29. Reid, G. C., *J. Geophys. Res.*, **66**, 4071 (1961).
30. Johnson, R. and Heppner, J. P., *Trans. Amer. Geophys. Union*, **37**, 350 (1956).
31. Page, F. M., *Disc. Faraday Soc.*, **37**, 203 (1964).
32. Elving, P. J., *Records Chem. Progr.*, **14**, 99 (1953).
33. Hedges, R. M. and Matsen, F. A., *J. Chem. Phys.*, **28**, 950 (1958).
34. Hollock, L. and Exner, H. J., *J. Chem. Phys.*, **56**, 46 (1952).
35. Lange, E., *Z. Electrochem.*, **56**, 94 (1952).
36. Matsen, F. A., *J. Chem. Phys.*, **24**, 602 (1956).
37. von Stackelberg, M. and Weber, P., *Z. Electrochem*, **56**, 806 (1952).

# The Experimental Determination
# of Electron Affinities

## 2.1 Early Work

The earliest attempts to evaluate electron affinities followed the development of the Born–Haber cycle[1,2] and were confined to a refinement of the calculation of lattice energies on which this cycle is based. The developments in this sphere, culminating with the very accurate calculations of Cubicciotti[3,4] on the alkali halide crystals have been reviewed by Waddington[5] and Ladd and Lee[6]. Other calculations including spectral extrapolations and quantum mechanical calculations culminating in Hylleraas extended analysis of $H^-$ [7] have proved fruitful in the field of atomic ions, and have been reviewed by Branscomb[8]. Apart from these calculations, the determination of the stabilities of gaseous negative ions proceeded slowly. The review by Pritchard[9], in 1953, lists only 64 ions, of which 6 had been determined by accurate calculations, 29 by similar calculations involving some empirical approximations of one form or another, 7 by studies on solutions, involving an estimate of solvation energies, and only 22 by experimental studies in the gaseous phase. Of these latter studies, which were made by a variety of techniques of widely differing reliability, few concerned ions other than the halogens. Since the review was published, several new techniques have become available, and others which received short mention then have been used in much wider fields.

## 2.2 Gaseous Equilibria

The direct study of the heat of a reaction such as

$$NaCl \rightleftharpoons Na^+ + Cl^-$$

can be used to give a value for the stability of the negative ion concerned. The greatest barrier to the widespread use of this approach is the difficulty of setting up the equilibrium, which becomes important only at high temperatures. The early work by Tandon[10] and others involved heating the salt vapour in a furnace and was limited to some alkali halides. Ionov, Dukelskii[11] and others used a heated filament to produce surface ionization, and, separating the products in a mass spectrometer, made comparative studies of several pairs of alkali halides. This work is discussed by Pritchard[9], and little has been added since that review. Again, the majority of the work concerned the halogens, although data on $H^-$, $CN^-$ and $S^-$ were also reported. A third method of establishing an equilibrium at high temperatures was through flame ionization. The electron concentration in a flame to which an alkali metal has been added in the form of a salt spray may be measured, either by microwave attenuation at atmospheric pressure, or by cyclotron resonance in a low pressure flame. The electron concentration may be modified by the addition of electron acceptors, and an analysis of the effect of such acceptors can give a value for the stability of the negative ion produced. The chemistry involved is, however, complex since the electron acceptor may interact with the alkali metal, or with the flame gases, and the analysis must extend over the whole distribution of products[12]. Furthermore, the addition of an electron acceptor may induce a completely different steady state[13], so that there is an apparent increase in electron concentration. It is not surprising that few results have been reported from this method, only $SH[14]$ and $CN[15]$ being added to the halogens. The extensive work on $OH[16]$ was later shown to be incorrect[17], as the fundamental observation was an artefact.

## 2.3 Spectroscopic Methods

Since the review by Pritchard[9] two new methods have been reported, of a very considerable accuracy. Both methods are basically spectroscopic determinations of the energy necessary to detach an electron. In the first, Branscomb and his co-workers have determined the onset of photo-detachment from various negative ions[18]. The negative ions were generated in a medium pressure discharge, and isolated by a mass filter. They then passed into an optical detachment chamber, where they crossed a high intensity light beam from a wide aperture monochromator (in the original work narrow band filters were used). If the wavelength of the light was short enough the electrons were detached and, being separated by a weak magnetic field, were recorded separately. By varying the wavelength of light used it was possible to determine the photodetachment spectrum, and hence the threshold for onset. Among the ions successfully studied by this technique were $I^-$ [18], $O^-$ [19,20,21], $OH^-$ [22,23], $S^-$ [24] and $C^-$ [25] while qualitative observations on $CN^-$, $CNO^-$ [26] and $NO_2^-$ [27] were reported. Some difficulty was experienced with the important ion $O_2^-$ [28,29] owing to a doubt about the form of the cross-section at threshold, and the reported affinity is at variance with other estimates[30].

The second method involved the direct photography of the absorption spectrum of the negative ion, and is due to R. S. Berry[31,32]. A small amount of alkali halide (again the published results are restricted to the halogens), was volatilized in a shock-tube, and the photodetachment continuum in the plasma created by the passage of the shock front was observed spectrographically and recorded photographically. The results obtained by this technique are in exact agreement with the photodetachment cross-section studies of Branscomb[18], and with the best lattice energy calculations on the alkali halides, due to Cubicciotti[3,4].

## 2.4 Electron Capture Studies

The advent of the electron capture detector for the gas chromatograph, introduced by Lovelock[33] for the detection of electronegative species, brought forward a fresh method for the study of the stability of negative ions. The exact mechanism by which the detector acts is as yet uncertain[34,35], but Becker and Wentworth[36] have applied it to a study of the stability of the negative ions of aromatic hydrocarbons. In outline, a source of $\beta$ particles (a tritium impregnated foil) creates a plasma of secondary electrons and positive ions within a cell through which a carrier gas (conveniently argon + 5% methane) flows. At intervals of about 100 μsec a 1 μsec pulse is applied to the cell of sufficient amplitude to collect all the electrons within the cell, and the quantity of electricity thus carried is measured. The presence of an electronegative gas, which can form negative ions, will lower the electron concentration and hence produce a lower reading. If the process is at equilibrium it is possible to evaluate the electron affinity either from one measurement, by statistical calculation, or from a series of measurements at different temperatures by thermodynamic arguments.

## 2.5 The Magnetron Technique

The technique which has been used most widely for the study of the stabilities of gaseous negative ions was the magnetron technique originally introduced by Mayer and his co-workers in 1935[37] to study the electron

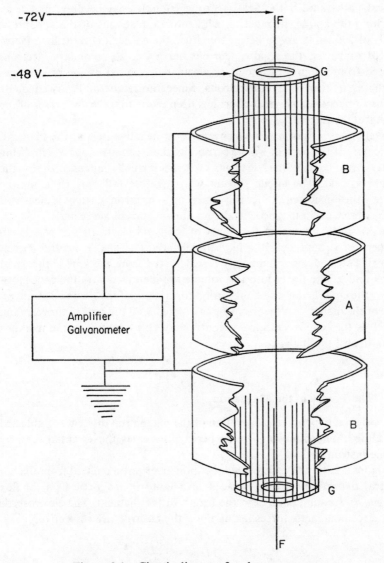

Figure 2.1  Circuit diagram for the magnetron

affinities of the halogens. In this technique, a hot filament emits electrons. In vacuo, these electrons are captured by a concentric anode held about 100 V positive. A concentric squirrel-cage grid is placed between the filament and anode, and is held at an intermediate potential (see Figure 2.1). The function of this grid is to sharpen the response of the apparatus. When a solenoidal magnetic field is applied the thermionic electrons describe helical paths and, if the field is strong enough, may be prevented from reaching the anode. In practice, with the latest designs of the apparatus, fields of 200 gauss are sufficient to reduce the electron current by a factor of $10^4$ or better, the residual current being due to secondary electrons ejected from the grid or anode, to electrons scattered by the residual gas in the system or to photoelectrons. Since the factor of $10^4$ is adequate for most purposes, no great effort has been made to trace the causes of the background current.

In the presence of a gas which can form negative ions at the filament, some part of the current will be carried by these negative ions which, being vastly heavier than the electrons, will be virtually undeflected by the magnetic field. The anode current will therefore fall only to a limiting value—the ion current. Since the total (ion + electron) current is measured in the absence of a magnetic field, and the ion current alone in its presence, it is possible to compare the fluxes of ions and electrons out of a small element of volume near the filament surface. The flux of neutral species into this element of volume may be calculated from the kinetic theory of gases and, if the fluxes into the volume are identified with the concentrations, the equilibrium constant for ion formation, at the assumed temperature of the filament surface, may be calculated, and the heat of the reaction, which is the required electron affinity, may be evaluated by the methods of statistical mechanics.

## 2.6   The Theory of the Magnetron

The original paper on the operation of the magnetron diode was published by Hull[38]. The discussion given here[39] illustrates the essential features of the instrument.

Figure 2.2 shows the forces acting upon an electron emitted from the hot central filament of a diode placed in a uniform magnetic field, of flux density $B$, running parallel to the length of the filament. The electrostatic and electromagnetic forces acting upon the electron are respectively:

$$F = -eE = -e\frac{dV}{dr}$$

(*F* is directed outward since the anode is positive) and $f = Bev$ where $e$ is the electronic charge, $V$ is the potential at a distance $r$ from the filament and $v$ is the velocity of the electron at this point. If $v$ is resolved into components $v_r$ and $v_\theta$, along and perpendicular to the radius vector, then the components of $f$ will be:

$$f_r = Bev_\theta \qquad \text{and} \qquad f_\theta = Bev_r$$

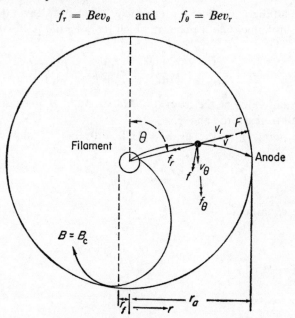

Figure 2.2   Motion of a charged particle within the magnetron

If $\theta$ is the angle between the radius vector, $r$, and an arbitrary line in the azimuthal plane, the angular velocity of the electron is

$$\omega = \frac{d\theta}{dt} = \frac{v_\theta}{r}$$

The equation of radial motion therefore becomes

$$\frac{d}{dt}\left(m\frac{dr}{dt}\right) = F - f_r = e\frac{dV}{dr} - Ber\frac{d\theta}{dt} \tag{2.1}$$

where $m$ is the electron mass.

The equation of azimuthal motion is found by equating the moment of the impressed force to the rate of change of the angular momentum, thus:

$$rf_\theta = rBe\frac{dr}{dt} = \frac{d}{dt}(mr^2\omega)$$

2+

Integrating each side with respect to time and noting that $\omega=0$ as the electron leaves the filament gives

$$\omega = \frac{Be}{2m} \left(1 - \frac{r_f^2}{r^2}\right) \tag{2.2}$$

If the small thermal velocity with which the electron leaves the filament is ignored, its velocity at a point at which its potential is $V$ will be given by

$$v = \sqrt{\frac{2e}{m} V} = \left[\left(\frac{dr}{dt}\right)^2 + r\left(\frac{d\theta}{dt}\right)^2 + \left(\frac{dz}{dt}\right)^2\right]^{\frac{1}{2}}$$

At a certain value of the magnetic field strength, $B = B_c$, the electron will just fail to reach the anode.

At this point, $V = V_a$; $dr/dt = dz/dt = 0$; $r = r_a$, and $d\theta/dt = \omega$, whence

$$\omega = \frac{v_\theta}{r_a} = \frac{1}{r_a} \sqrt{\frac{2e}{m} V_a} \tag{2.3}$$

Therefore, equating (2.2) and (2.3)

$$B_c = \frac{\sqrt{8V_a}}{r_a[1 - (r_f^2/r_a^2)]\sqrt{e/m}} \tag{2.4}$$

whence, if $r_a \gg r_f$,

$$B_c = \sqrt{\frac{8mV_a}{er_a^2}} \tag{2.5}$$

Equation (2.4) shows the value of $B_c$ to be independent of the potential distribution between the anode and cathode so that the presence of other electrodes or space charge effects should not alter the cut-off conditions. In practice the cut-off is not so sharp with grids present, probably because of the distortion of the magnetic and electric fields by these structures. The efficiency of the removal of electrons was, however, much higher.

In the magnetron triode the electrons are captured by the grid and so prevented from creating a large space charge near the filament which would affect the emission of the negative ions. The magnetic field strength is usually made greater than $B_c$ so that the apogee occurs near the grid. This causes the electrons to approach it tangentially and results in a greater efficiency of electron removal. The presence of a second grid was shown by Page[40] to lead to a further increase in efficiency, presumably because of its action in trapping electrons which had passed through the first grid because of potential and geometrical asymmetry within the apparatus.

## 2.7 The Design of the Apparatus

Figure 2.3 shows two designs of the magnetron which have been widely used in this work. The early design (*a*) suffered from a lack of rigidity and other faults, and was replaced by an improved design (*b*). The apparatus consists essentially of a central filament (F) surrounded by two coaxial grids ($G_1$, $G_2$) and an anode (A) which is flanked by guard plates ($P_1$, $P_2$). These latter ensure an even temperature over the length of filament from which the measured thermionic currents derive. In the early design[40] the anode and guard plates were formed from a layer of vacuum-deposited silver, the plates being separated (J) by marking with a steel scribe. When using this system, considerable difficulty was encountered in ensuring that the silver formed a good electrical contact with the tungsten seals set in the wall of the bottle. The main disadvantage of the system, however, seemed to result from the adsorption of vapours upon the scribed marks, rapidly leading to electrical short-circuiting between the anode and the plates. It proved to be impossible to clean the apparatus adequately after this had occurred without dissolving off the silver and replating. This resulted in the bottle becoming badly marked in the region J. The ground glass joint between the top cap and the main body of the bottle also tended to seat very firmly and since the grid assembly was rather delicate, it was easily broken while being dismantled. A further disadvantage of this arrangement was a lack of rigidity in the filament and grid supports making the apparatus very sensitive to vibration.

In the final design, the rigidity was improved by mounting the grids upon 6 B.A. brass studding which was insulated by inserting it into glass tubing which also served to add some support. The plated anodes were replaced by 4 cm long by 1 mm thick molybdenum sheet which was sprung into position. The connexions to this were made by inserting copper wire between the sheet and the wall of the bottle. The connexion to the tungsten seals P and A were made by means of miniature 'crocodile' clips. This arrangement made it a simple matter to remove the anode and plates and clean them and the bottle without disturbing the filament assembly.

In the early apparatus the filament was secured at one end by a grub screw and at the other by means of a tensioning spring (S) made of spring steel. This arrangement worked well for tungsten filaments, but was incapable of retaining sufficient tension in more ductile filaments without stretching them and causing them to sag into the grids when hot. This was overcome by making the spring out of 0·01in. diameter tungsten wire. This proved to be ideal for the purpose and enabled a much wider range of metals to be used as filament materials. The filament assembly was

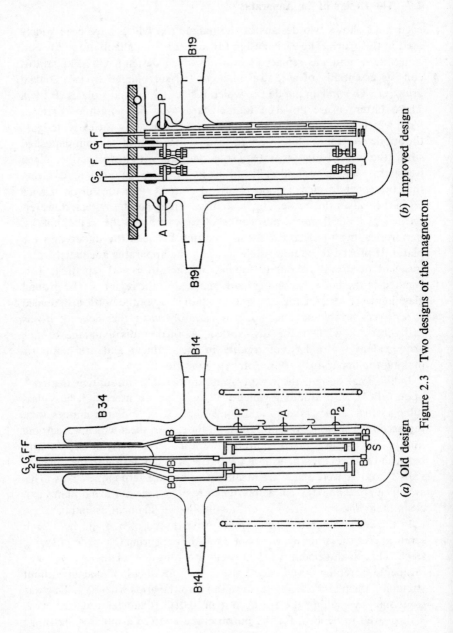

(a) Old design                    (b) Improved design

Figure 2.3   Two designs of the magnetron

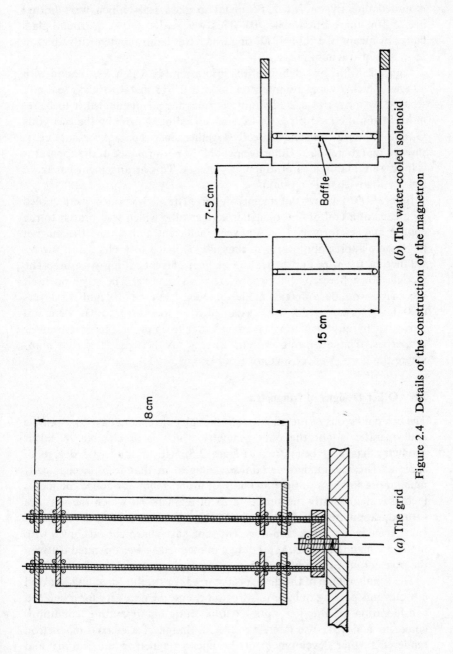

(b) The water-cooled solenoid

(a) The grid

Figure 2.4   Details of the construction of the magnetron

mounted directly on A.E.I. S5 metal to glass seals which were brazed
into a $\frac{1}{4}$in. thick brass-plate (B). This was sealed on to a ground-glass
flange by means of a rubber 'O' ring which was lightly coated with Apiezon
L or M high-vacuum grease.

Figure 2.4(*a*) shows details of the grid assembly which was wound with
40 s.w.g. nickel wire, mounted on an A.E.I. S13 metal to glass seal and
secured by means of a 2 B.A. nut. This arrangement enabled it to be removed from the base-plate for ease of rewinding. Normally the two grids
$G_1$ and $G_2$ were connected directly together since separate measurements
showed that there was no improvement in the performance of the apparatus
if they were maintained at different voltages. This arrangement has more
rigidity than separate mountings.

Figure 2.4(*b*) shows the magnet design. The early models were cooled
by means of a coil of $\frac{1}{4}$in. diameter copper tube which was brazed to the
copper magnet former. This was very inefficient and caused the magnet
to become appreciably warm during use. This led to a change in the re-
sistance of the wire and hence the magnet current falling with use. The
model shown here completely obviated this trouble. The solenoid itself
was wire wound, different models having between 10 and 15 layers
(900–1500 turns) of 20 s.w.g. enamelled copper wire which were not
separated by any insulating layers in order to improve the heat transfer
properties. The operating current usually lay between 4 and 6 amps
where the heating effect was not troublesome.

## 2.8   Other Designs of Magnetron

The coaxial layout of the magnetron, though convenient, easy to assemble
and operate, is not the only geometry which is practicable. A linear
geometry has also been tried (Figure 2.5) with a moderate degree of
success. This design has certain advantages, in that it is mechanically
much more stable, and the possibility of using ribbon, or coiled, filaments
increases significantly the surface area of filament on which the ions are
formed, despite the smaller angle subtended by the anode at the filament.
Furthermore, the long, field-free drift tube gave sharp cut-offs, even with
moderate magnetic fields. (This design of magnetron was operated between
the poles of an electromagnet.)

The disadvantage of the design is the need for careful screening to avoid
any currents passing outside the drift tube, and the necessity for a detailed
consideration of the ion optics before optimum operating conditions
could be achieved. The later, more rigid, design of a coaxial magnetron
rendered further development of the linear magnetron unnecessary and
though it was used in some experiments it has since been abandoned.

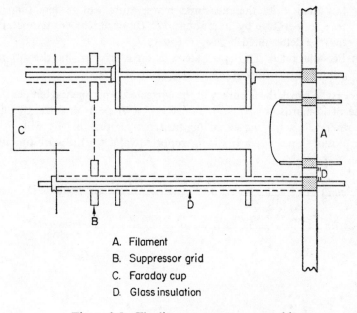

A. Filament
B. Suppressor grid
C. Faraday cup
D. Glass insulation

Figure 2.5   The linear magnetron assembly

## 2.9   The Measurement of Filament Temperature

The measurement of the temperature of the filament is fundamental to
the determination of energies by the magnetron technique. If the methods
of statistical mechanics are used to evaluate energies from the data on ion
and electron currents it is, of course, desirable that this data be known
fairly accurately, and absolutely, but it is of prime importance that the
temperature be known accurately. If the data is to be analysed by the
arguments of thermodynamics, then the accuracy demanded of the data is
increased, but it is enough that this data, and the temperatures, be correct
relative to the other data in the set, the absolute values being of lesser
importance.

The usual method of determining the temperature of the filament has
been by the use of a disappearing filament pyrometer, measuring through
one thickness of glass. The pyrometer reading will always be lower than
the true filament temperature because of:

(i) Absorption by the glass. In the temperature range 1000–2000°K
    the mean correction for absorption by the glass is about $+24°K$[41].
(ii) Emissivity of the surface. When a filament material has an emissivity

less than unity the measured temperature will be less than the true temperature by an amount $\Delta T$. The true filament temperature may be determined by use of Figure 2.6.

(iii) Personal error. This is a systematic error arising from the differing colour sensitivities of different operators.

It is easy to check the accuracy of the corrected experimental temperature by use of Richardson's equation, since a plot of electron current against $1/T_f$, where $T_f$ is the measured filament temperature in °K, will give a linear plot of slope $-\chi/R$. If $\chi$ is the reported value for the work function,

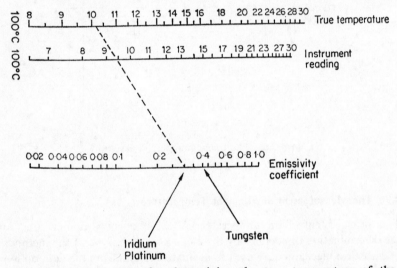

Figure 2.6  Nomogram for determining the true temperature of the filament

the true temperature, $T$, can be calculated since, assuming that $T_f = T + \delta$,

$$\frac{\mathrm{d}\log j_e}{\mathrm{d}(1/T_f)} = \frac{\mathrm{d}\log j_e}{\mathrm{d}(1/T)}\frac{\mathrm{d}(1/T)}{\mathrm{d}(1/T_f)} = \left(\frac{T}{T+\delta}\right)^2\frac{\mathrm{d}\log j_e}{\mathrm{d}(1/T)} \qquad (2.6)$$

Hence

$$\chi_1 = \chi\left(\frac{T}{T+\delta}\right)^2$$

or

$$\frac{\delta}{T} = \left(\frac{\chi}{\chi_1}\right)^{\frac{1}{2}} - 1 \qquad (2.7)$$

The correction $\delta$ will include all three systematic errors, unless those due to absorption or emissivity have been allowed for before measuring the apparent work function.

Alternatively, the true temperature may be calculated directly from Richardson's equation, provided the thermionic current density is known.

$$j_e = BAT^2 \exp\left(\frac{-\chi}{RT}\right)$$

where $B$ is a constant equal to 120 amp cm$^{-2}$ deg$^{-2}$

    $A$ is surface area cm$^2$

    $T$ is the true temperature °K

    $\chi$ is the work function of the clean surface in kcal mole$^{-1}$.

Again, all three errors are allowed for unless preliminary corrections have been made.

In order to determine and apply these corrections it is usual to smooth the direct observations graphically before commencing the analysis, to remove the random errors of measurement. This smoothing may be done by plotting $T_f$ against $j_f$ and drawing a curve through the points, but the curvature is quite sharp, and any interpolation, particularly at low temperatures, is not very reliable. A better method of carrying out the graphical smoothing is to plot $j_f$ against $T_f^2$. This is based on Stefan's law

$$E = \sigma T^4$$

If the heat loss by conduction at the ends of the filament is neglected the heat supplied may be set equal to the total radiation emitted, $E$.

i.e. $$E = j_f^2 R = \sigma T^4$$

or $$j_f = kT^2 \tag{2.8}$$

Hence a plot of $j_f$ against $T^2$ is linear, making extrapolations to very low and very high temperatures more reliable.

As some workers have found difficulty in using the optical pyrometer for temperature measurements on thin filaments a method using the electron emission directly has been developed[42]. This method has also been found useful under circumstances where continuous deposition on, or etching of, the filament occurs.

If the electron current ($j_e$) is given by Richardson's equation, then this equation may be rewritten as

$$\log j_e = A - \frac{\chi}{RT} \tag{2.9}$$

or $$\frac{1}{T} = \frac{AR}{\chi} - \frac{R}{\chi}\log j_e$$

where $\chi$ = work function eV

    $T$ = temperature °K.

2*

The general expression used to evaluate the apparent electron affinity ($E'$) from the ion ($j_i$) and electron ($j_e$) currents at a pressure $p$ is

$$\log\left(\frac{j_e p}{j_i}\right) = B - \frac{E'}{RT}$$

$$= B - \frac{E'}{R}\left(\frac{AR}{\chi} - \frac{R}{\chi}\log j_e\right) \quad (2.10)$$

$$= C + \frac{E'}{\chi}\log j_e \quad (2.11)$$

so that a plot of $\log(j_e p/j_i)$ against $\log j_e$ yields a straight line of slope $(+E'/\chi)$.

Alternatively, it may be shown that a plot of $\log(j_i/p)$ against $\log j_e$ is a straight line of slope $[(\chi - E)/\chi]$.

## 2.10  Thermionic Work Functions

It is implied in the foregoing paragraphs that the measurement of thermionic work functions in the magnetron is valid and reliable, and a considerable portion of Chapter 5 will be devoted to a discussion of the effect of adsorbed gases on these work functions. Throughout the course of the measurements discussed in these pages, it has always been possible, after strong pumping and heating, to reach a reproducible state, with a clearly defined temperature dependence of the electron current.

This temperature dependence, although reproducible, could not be held to be a fundamental property of the filament material, since it showed a tendency to vary from one batch to another, and to show a slow ageing, but it was nevertheless a well-defined and readily measurable quantity,

Table 2.1  Practical Work Functions

| | Experimental[42,43] | | Literature[44,45] |
|---|---|---|---|
| | eV | kcal mole$^{-1}$ | eV |
| Tungsten | 4·56 | 105 | 4·5 |
| Tantalum | 4·21 | 97 | 4·12 |
| Platinum | 4·78 | 110 | 5·32 |
| Iridium Sample A | 3·83 | 88 | } 5·3 |
| Sample B | 4·44 | 102 | |
| Rhenium | 4·0 | 92 | 5·0 |
| Tungsten Carbide | 4·85 | 112 | 4·58 |
| Tantalum Carbide | 5·24 | 121 | 3·77 |
| Molybdenum Carbide | 4·80 | 111 | 3·85 |
| Carbon | 4·69 | 108 | 4·7 |

All results $\pm 0\cdot 2$ volt.

and one which could be used as the work function of a particular filament with confidence. Usually the variations were slight, being within the experimental error. These experimental work functions were close to the values quoted in the literature, and could be used as a test of the proper functioning of the apparatus.

The practical work functions for commonly used filament materials are set out in Table 2.1.

## REFERENCES

1. Born, M., *Ber. Deut. Phys. Ges.*, **21**, 13 (1919).
2. Born, M. and Mayer, J. E., *Z. Physik*, **75**, 1 (1932).
3. Cubicciotti, D. D., *J. Chem. Phys.*, **31**, 1646 (1959).
4. Cubicciotti, D. D., *J. Chem. Phys.*, **33**, 1579 (1960).
5. Waddington, T. C., *Advan. Inorg. Chem. Radiochem.*, **1**, 157 (1959).
6. Ladd, M. F. C. and Lee, W. H., *Progress in Solid-State Chemistry*, Vol. 1, Pergamon Press, Oxford, 1964.
7. Hylleraas, E. A., *Astro. Phys. J.*, **111**, 209 (1950).
8. Branscomb, L. M., Chapt. 4, *Photodetachment*, in *Atomic and Molecular Processes* (Ed. D. R. Bates), Academic Press, New York, 1962.
9. Pritchard, H. O., *Chem. Rev.*, **52**, 529 (1953).
10. Tandon, A. N., *Proc. Natl. Acad. Sci. India*, **7**, 102 (1937).
11. Ionov, N. I. and Dukelskii, V. M., *J. Exptl. Theoret. Phys. USSR*, **10**, 1248 (1940).
12. Page, F. M., *Ph.D. Thesis*, Cambridge, 1955.
13. Padley, P. J., Page, F. M. and Sugden, T. M., *Trans. Faraday Soc.*, **57**, 1552 (1961).
14. Page, F. M., *Rev. Inst. Français Petrole*, **13**, 692 (1958).
15. Padley, P. J., *Private communication*.
16. Page, F. M., *Disc. Faraday Soc.*, **19**, 87 (1955).
17. Kay, J. and Page, F. M., *Trans. Faraday Soc.*, **62**, 3081 (1966).
18. Branscomb, L. M., Seman, M. L. and Steiner, B., *J. Chem. Phys.*, **37**, 1200 (1962).
19. Branscomb, L. M. and Smith, S. J., *J. Res. Nat. Bur. Stds.*, **55**, 165 (1955).
20. Branscomb, L. M. and Smith, S. J., *Phys. Rev.*, **98**, 1127 (1955).
21. Branscomb, L. M., Burch, D. S., Smith, S. J. and Geltman, S., *Phys. Rev.*, **111**, 503 (1958).
22. Branscomb, L. M., *Advan. Electron. Electron Phys.*, **9**, 43 (1957).
23. Smith, S. J. and Branscomb, L. M., *Phys. Rev.*, **99**, 1657 (1955).
24. Branscomb, L. M. and Smith, S. J., *J. Chem. Phys.*, **25**, 598 (1956).
25. Branscomb, L. M. and Seman, M. L., *Phys. Rev.*, **125**, 1602 (1962).
26. Smith, S. J., *Proc. 4th Intern. Conf. Ionisation Phenomena*, Uppsala, 1959, North Holland, Amsterdam, 1960.
27. Branscomb, L. M., *Threshold of Space*, p. 101, Pergamon Press, New York, 1957.
28. Burch, D. S., Smith, S. J. and Branscomb, L. M., *Phys. Rev.*, **112**, 171 (1958).
29. Burch, D. S., Smith, S. J. and Branscomb, L. M., *Phys. Rev.*, **114**, 1652 (1959).

30. Branscomb, L. M., Chapt. 4, *Photodetachment*, in *Atomic and Molecular Processes* (Ed. D. R. Bates) p. 130, Academic Press, New York, 1962.
31. Berry, R. S., Reimann, C. W. and Spokes, G. N., *J. Chem. Phys.*, **35**, 2237 (1961).
32. Berry, R. S., Reimann, C. W. and Spokes, G. N., *J. Chem. Phys.*, **37**, 2278 (1962).
33. Lovelock, J. E., *Anal. Chem.*, **33**, 162 (1961).
34. Wentworth, W. E., Chen, E. and Lovelock, J. E., *J. Phys. Chem.*, **70**, 445 (1966).
35. Lyons, L. E., Morris, G. C. and Warren, L. J., *Australian J. Chem.*, **21** 853 (1968).
36. Becker, R. S. and Wentworth, W. E., *J. Am. Chem. Soc.*, **85**, 2210 (1963).
37. Mayer, J. E. and Sutton, P. P., *J. Chem. Phys.*, **3**, 20 (1935).
38. Hull, A. W., *Phys. Rev.*, **18**, 31 (1921).
39. Parker, P., *Electronics*, Arnold, London, 1963.
40. Page, F. M., *Trans. Faraday Soc.*, **56**, 1742 (1960).
41. Parker, P., *Electronics*, Appendix 9, p. 1014, Arnold, London, 1963.
42. Gaines, A. F. and Page, F. M., *Trans. Faraday Soc.*, **59**, 1266 (1963).
43. Farragher, A. L., *Ph.D. Thesis*, The University of Aston in Birmingham, 1966.
44. Fomenko, V. S., *Handbook of Thermionic Properties*, Plenum Press Data Division, New York, 1966.
45. Reimann, A. L., *Thermionic Emission*, Chapman and Hall, London, 1934.

# 3

# The Classification of Results

## 3.1 The Effects of Filament Temperature

When the temperature of the filament in the magnetron is increased both the electron and ion currents are observed to increase simultaneously, though rarely at equal rates. Mayer[1] related the two currents by considering the fluxes of ions, electrons and neutral species into a small element of volume near the filament surface. This treatment worked well for negative ions in whose formation the surface either played no part, or acted only as an inert source of electrons. The results outlined in Chapter 5, where the adsorption of a fragment of the neutral molecule on the filament surface contributes to the observed energetics, cannot be accounted for by this simple approach, and various, more sophisticated, theoretical treatments have been developed to take such surface reactions into account[2].

The basic experimental observations on which any theoretical treatment must be based are four in number. Apart from the demonstrable effect of adsorption, it may be shown that both the ion and electron currents are exponentially dependent on the reciprocal of the filament temperature, and that the ion current is usually proportional to the pressure of vapour in the magnetron. From this it is implicit that the ratio $j_e p / j_i$, where $j_e$ and $j_i$ are the electron and ion currents, and $p$ is the pressure, may be identified with some equilibrium constant, whose temperature coefficient leads to an energy term, which is the heat of some hypothetical reaction. In fact the energy term is the overall enthalpy change between the gas phase neutral species and the resultant negative ion.

## 3.2  The Simple Energetic Treatment

Accepting the identification of the observed energy as representing the overall energy change, the initial and final states may be defined. The initial state is an electron in the metal of the filament and a gaseous neutral species, and the final state that of a gaseous negative ion with, in certain cases, some fragment of the neutral species adsorbed on the metal surface, or free in the gas phase. The mechanism by which the transition from initial to final state is achieved involves chemisorption, ion formation on the surface, desorption of the ion, and in certain cases desorption of uncharged fragments; and the energetics of none of these steps are known. The overall energy change ($W$) may be written down, from Figure 3.1, as

$$W = \chi + D - E_A - Q_B \tag{3.1}$$

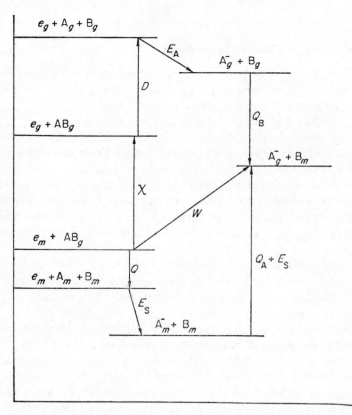

Figure 3.1   Energy level diagram for ion formation at a surface

where $\chi$ is the electron work function of the filament, $D$ the appropriate bond dissociation energy in the molecule, $E_A$ the electron affinity of the fragment A and $Q_B$ the heat of adsorption of fragment B.

$W$ will be the overall energy of ion formation, and since the ion current is usually compared with the electron current, the observed energy, $E'$, will be

$$E' = \chi - W$$
$$= E_A + Q_B - D \tag{3.2}$$

## 3.3 Kinetic Treatment of Ion Formation

A simple kinetic treatment of the processes of ion formation may be based on the mechanism suggested in the previous paragraph[3].

The ion current ($j_i$) may be identified with the rate of emission of negative ions from the surface, and hence, with their rate of desorption from the surface. If we assume that any ion precursor A may be desorbed (emitted) as a negative ion, as well as a neutral species, the respective heats of desorption being $Q_n$ and $Q_m$, and that a fraction $\theta$ of the surface is covered with adsorbed A, then, for unit area of surface,

$$j_i = C\theta \exp\left(\frac{-Q_n}{RT}\right) \tag{3.3}$$

The electron current is given by Richardson's equation, but the fraction of the surface which may emit electrons is only $(1 - \theta)$

$$j_e = B(1 - \theta)T^2 \exp\left(\frac{-\chi}{RT}\right) \tag{3.4}$$

The fraction of surface covered, $\theta$, may be calculated from Langmuir's equation

$$\frac{\theta}{1 - \theta} = \frac{k_1 p}{k_2} = \frac{k_1 p}{K}\left[\exp\left(\frac{+Q}{RT}\right)\right] \tag{3.5}$$

where $Q$ is the heat of adsorption of the neutral species AB, and $p$ the pressure.

Hence
$$\frac{j_e p}{j_i} = \frac{[BT^2(1 - \theta)\exp(-\chi/RT)]p}{[C\theta\exp(-Q_n/RT)]}$$

$$= \frac{BT^2 p}{C}\frac{(1 - \theta)}{\theta}\exp\left[\frac{-(\chi - Q_n)}{RT}\right]$$

$$= \frac{BT^2 K}{Ck_1}\exp\left[\frac{-(\chi - Q_n + Q)}{RT}\right] \tag{3.6}$$

$Q$, the heat of adsorption of AB, is related to the energy for dissociation into A and B (D) and the heats of adsorption of A and B ($Q_A$, $Q_B$) by

$$Q = -D + Q_A + Q_B \tag{3.7}$$

while $Q_n$, the heat of desorption as a negative ion, may be equated to the work done in desorbing a neutral A and an electron, and combining them in the gas phase

$$Q_n = Q_A + \chi - E \tag{3.8}$$

Hence

$$\frac{j_e p}{j_i} = \frac{BT^2 K}{C k_1} \exp\left[\frac{-(E + Q_B - D)}{RT}\right] \tag{3.9}$$

## 3.4  The Mechanism of Ion Formation

The constants $K$, $C$ and $k_1$ of the previous section may actually show some slight dependence on temperature, so that the graphical evaluation of the energy term, by plotting $\log(j_e p/j_i)$ against $1/T$, will lead to a straight line whose slope, the apparent electron affinity $E'$, is given by

$$E' = E + Q_B - D + nRT \tag{3.10}$$

The value of $n$ is discussed in Chapter 6 in connexion with the correction of the observed values at $0°K$. For the present, the value of $n$ will be taken as zero.

It does not follow that every negative ion is formed by a process involving the adsorption of a residue B, and four principal types of mechanism may be distinguished on the grounds of the energy terms which contribute to the apparent electron affinity.

### Type I  Direct Capture

Certain neutral species have been found to add an electron, to form a negative ion. This ion will always be a doublet molecule (anionic free radical) unless the parent neutral molecule was a stable free radical. One of the few such free radicals is $NO_2$, and the resultant singlet ion is the nitrite ion ($NO_2^-$)[4]. More usually the direct acceptor is a compound with either many constituent electronegative elements, e.g. sulphur hexafluoride[5]

$$SF_6 + e \rightarrow SF_6^-$$

or an extended $\pi$ bond system, e.g. fluorobenzoquinone[6]

$$C_6H_3FO_2 + e \rightarrow C_6H_3FO_2^-$$

Since the process of direct capture involves neither bond breaking nor adsorption, the apparent electron affinity must be equal to the true electron affinity at the temperature of measurement.

## Type II   Weak Bond

The transition between a stable free radical where direct capture occurs and a stable molecule wherein a bond of high energy has to be broken, is a gradual one, and no sharp distinction can be drawn. Generally speaking, if free radicals can capture an electron to form a negative ion, then, if the strength of the bond which has to be broken to form the free radicals is less than about 70 kcal mole$^{-1}$, the substance will behave as if it were composed of the free radical, and the apparent electron affinity and true electron affinity will be the same. This type of behaviour is shown by two such classes of molecules.

IIa   Symmetrical fission
e.g. Hydrazine[7]

$$N_2H_4 (\to 2NH_2) + 2e \to 2NH_2^-$$

IIb   Unsymmetrical fission
e.g. Dimethylmercury[8]

$$(CH_3)_2Hg (\to 2CH_3 + Hg) + 2e \to 2CH_3^- + Hg$$

In enquiring why it should ever be possible to omit the bond energy term, a model may be constructed in which an equilibrium between the neutral, singlet, molecule and the free radicals is set up near the filament. If the equilibrium lies towards the side of the free radicals so that the degree of dissociation is over 50%, increasing the temperature will have only a slight effect on the number of radicals reaching the filament, so that no significant error is introduced in assuming that dissociation is complete, that is, in neglecting the bond energy. If the degree of dissociation is low, altering the temperature alters the number of free radicals reaching the filament, so that the bond energy must be taken into account. This hypothesis of a pre-equilibrium has many drawbacks, but it does offer an explanation of the arbitrary way in which compounds appear to follow this, or the next type of mechanism. The equilibrium constant is determined not only by the energy but also by the entropy of the reaction and as the surface may be involved, this may vary considerably. A favourable entropy may induce type II behaviour even with a high bond energy whereas an unfavourable entropy may force type III behaviour with a low bond energy. It has, on occasion, been possible to upset the dissociation equilibrium by altering the external conditions. For example, dibenzyl normally shows a type III behaviour, but if the vapour enters the vessel over a silver deflector plate heated electrically to about 150°C, type II behaviour is consistently shown[9].

*Type III   Strong Bond*

This type of behaviour is observed when the bond energy is so high that only a small fraction of the material is dissociated at the filament temperature. It is a more general type of behaviour than type II, since the energy of every step in the mechanism contributes to the total, and the apparent electron affinity is the true electron affinity, less the work necessary to create the acceptor radicals. As with type II, two sub-classes may be distinguished.

III*a*   Symmetrical fission

$$E' = E - D/2$$

e.g. Oxygen[3]

$$O_2 (\rightarrow 2O) + e \rightarrow 2O^- [D \equiv D(O\text{---}O)]$$

III*b*   Unsymmetrical fission

$$E' = E - D$$

e.g. Carbon tetrachloride[10]

$$CCl_4 (\rightarrow CCl_3 + Cl) + e \rightarrow CCl_3 + Cl^- [D \equiv D(CCl_3\text{---}Cl)]$$

*Type IV   Dissociation with Adsorption*

This mechanism has been referred to already. Energetically, both a bond dissociation energy and a heat of adsorption contribute to the observed result, and the apparent electron affinity is given by

$$E' = E + Q - D$$

Many examples might be quoted, showing adsorption of different residues.

Carbon tetrachloride[10]

$$CCl_4 + e \rightarrow CCl_3^- + Cl_{ads} \begin{bmatrix} D \equiv D(CCl_3\text{---}Cl) \\ Q \equiv Q(Cl/Pt) \end{bmatrix}$$

Benzene[11]

$$C_6H_6 + e \rightarrow C_6H_5^- + H_{ads} \begin{bmatrix} D \equiv D(C_6H_5\text{---}H) \\ Q \equiv Q(H/WC) \end{bmatrix}$$

It is possible for other energies to enter into the total, if the resultant negative ion rearranges. For example, azulene forms a negative ion[12]

$$C_{10}H_8 + e \rightarrow C_{10}H_7^- + H_{ads}$$

but the energetics observed for this process imply an unreasonably high electron affinity for the azulyl radical. However, azulene is less stable than its isomer, naphthalene, by 30 kcal mole$^{-1}$ and if the negative ion had isomerized to the naphthyl ion, the experimental electron affinity would be in good agreement with expectation.

### 3.5 The Diagnosis of Reaction Type

A substance which can give rise to negative ions in the magnetron may usually do so in several different ways. Plausible reaction schemes may be set up leading to different negative ions, or alternative routes devised to produce one ion. Experimentally, a single substance may show different apparent electron affinities in different temperature ranges and, if any useful deductions are to be made, the processes giving rise to the negative ions in each range must be identified. Occasionally it is possible to do this unequivocally, as all the energies are known, but more usually several alternative processes have to be considered.

A mass spectrometer was used by Rosenstock and Herron to observe the ions directly[13]. They were able to show that tetracyanoethylene and di-*t*-butyl peroxide produced the expected negative ions by types I and II processes, but that benzene and sulphur hexafluoride produced a variety of negative ions whose relative proportions changed with the temperature. These observations were too limited to provide a basis for general diagnosis of reaction mechanism, and are being extended in these laboratories using a quadrupole mass filter. If such a method of identification proves to be generally possible, the conclusions reached will be greatly strengthened. Those results of Rosenstock and Herron which showed the greatest variety of ions were also those which had proved most erratic and difficult to study in the magnetron, and where a long conditioning of the filament was needed before reproducible results were obtained. It is therefore most important that the mass spectrometry and magnetron measurements be carried out under similar conditions if valid conclusions are to be drawn.

The possibility that the magnetron measurements refer to metastable ions must not be overlooked. If the ion which is formed, and whose energetics are determined in the magnetron, is metastable in that it changes to another negative ion before it is detected, the observed energy will depend on the primary ion, but the mass spectrometer will record only the secondary ion.

Even though mass spectrometry is potentially such a powerful tool, very little use of it has as yet been made, and diagnosis of reaction type is usually based on three considerations: the magnitude of the apparent electron affinity, the effect of varying the filament, and the calculation of the entropy of the reaction.

## 3.6 Considerations of Energy

It is immediately apparent from section 3.4, that whilst any type of mechanism may lead to a positive value of the apparent electron affinity, only types III and IV, where a bond is broken, can lead to the negative values which are sometimes observed. As a corollary to this, it is unlikely, though not impossible, that a bond can be broken if a large and positive apparent electron affinity is found, so that an observed energy of $+50$ kcal mole$^{-1}$ or greater implies a type I or II process.

These limits may be defined a little more closely if the entropies of ion formation are considered (section 3.8) in conjunction with the range of observations possible in the magnetron. The lower limit for the measurement of an ion current is when this becomes comparable with the background current, while an upper limit is set by the electron current being small compared with the ion current. The observable range in the magnetron may be defined by the inequality

$$10^{-6} < \frac{j_i}{j_e} < 10 \tag{3.11}$$

Equation (3.15) may be rewritten at typical operating conditions of $1200°K$ and $10^{-3}$ mm Hg (these values are not critical) as

$$\frac{E'}{T} = \Delta S + R \ln \frac{j_i}{j_e} - 59 \cdot 4 \tag{3.12}$$

The lowest value of $E'/T$ for a type I process ($\Delta S = 110$ cal deg$^{-1}$ mole$^{-1}$) is therefore 23, and the greatest value for a type IV process ($\Delta S = 82$ cal deg$^{-1}$ mole$^{-1}$) is 17 while a type III process ($\Delta S = 53$ cal deg$^{-1}$ mole$^{-1}$) must always have a negative $E'$.

The lowest temperature which can be used is set by the current ratio and Richardson's equation for thermionic emission, and for the type I process is $1200°K$, so that the lowest apparent electron affinity to be expected from this process is 28 kcal mole$^{-1}$. The highest temperatures commonly reached are about $2000°K$ so that the highest apparent electron affinity likely to be associated with a type III process will be 34 kcal mole$^{-1}$.

A consideration of the experimental energy of ion formation can therefore give a useful indication of the mechanism by which the ions are formed. If it is negative, the process is probably type III; if it lies between

Table 3.1  Possible Products from Fluoroform

| Type | Products | Electron Affinity | Notes | |
|---|---|---|---|---|
| a | I | $CHF_3^-$ | 8·6 kcal mole⁻¹ | $E_0 = E'$ | |
| b | III | $CF_3^- + H$ | 111·6 kcal mole⁻¹ | $E_0 = E' + D$ | $D = 103^{15}$ kcal mole⁻¹ |
| c | III | $CF_3 + H^-$ | 111·6 kcal mole⁻¹ | $E_0 = E' + D$ | $D = 103^{15}$ kcal mole⁻¹ |
| d | III | $CHF_2 + F^-$ | 129·6 kcal mole⁻¹ | $E_0 = E' + D$ | $D = 121$ kcal mole⁻¹ |
| e | III | $CHF_2^- + F$ | 129·6 kcal mole⁻¹ | $E_0 = E' + D$ | $D = 121$ kcal mole⁻¹ |
| f | IV | $CF_3^- + H_{ads}$ | 42·6 kcal mole⁻¹ | $E_0 = E' + D - Q$ | $Q = 69$ kcal mole⁻¹ |
| g | IV | $CHF_2^- + F_{ads}$ | 79·6 kcal mole⁻¹ | $E_0 = E' + D - Q$ | $Q = 50$ kcal mole⁻¹ |
| h | IV | $H^- + CF_{3ads}$ | 19  kcal mole⁻¹ | $E_0 = E' + D - Q$ | $Q = 92·6$ kcal mole⁻¹ |
| i | IV | $F^- + CHF_{2ads}$ | 80  kcal mole⁻¹ | $E_0 = E' + D - Q$ | $Q = 49·6$ kcal mole⁻¹ |

0 and 35 kcal, type IV; and above this type I (or II). These ranges may overlap, but not by more than ten kilocalories.

As an example of the arguments used, fluoroform ($CHF_3$) may be considered. This substance, when examined over a platinum filament[14], showed an apparent electron affinity of $16.7 \pm 1.8$ at a mean temperature of $1342°K$, which corrected to $0°K$ became $8.6 \pm 1.8$ kcal mole$^{-1}$.

The C—H and C—F bonds in this molecule are strong (103 and 121 kcal mole$^{-1}$ [15] respectively) so that no type II process is possible. One type I, four type III and four type IV processes may, however, be written down.

(*a*) is unlikely, as the electron affinity of $8.6$ kcal mole$^{-1}$ is far too low to be observed directly in the magnetron.

(*b*) to (*e*) yield impossibly high electron affinities, and in (*c*) and (*d*) the values disagree with the accepted value for the particular ion. The electron affinity predicted by (*g*) is improbably high when compared with similar fragments, and the same is true of the heat of adsorption predicted from (*h*). The only two processes which are at all probable are the type IV processes (*f*) and (*i*), and the former is favoured, since hydrogen atom adsorption is a well-established process, while the adsorption of $CHF_2$, a tetraatomic radical, is unprecedented, though the predicted heat of adsorption is not unreasonable.

The final argument which may be used to confirm (*f*) as the mechanism of ion formation in fluoroform is to examine other compounds which could give rise to $CF_3^-$ to show that a consistent value for the electron affinity of $CF_3$ is obtained.

## 3.7   Variation of the Filament Material

If a consideration of the energy indicates that a type IV process is occurring, it is possible to obtain confirmatory evidence by changing the material used as filament. The heats of adsorption of hydrogen atoms on the commonly used filament materials, tungsten and iridium, are 72 and 66 kcal mole$^{-1}$ respectively, and the difference, 6 kcal mole$^{-1}$, is larger than the experimental uncertainty usually found. A comparison of the apparent electron affinities measured with different filaments can confirm that a type IV process is occurring if a difference is found, or deny it if there is little difference.

The apparent electron affinity of benzene, corrected to $0°K$, was $25.1$ kcal mole$^{-1}$ on a tungsten (probably tungsten carbide) filament[11], while on an iridium filament it was $20.7$ kcal mole$^{-1}$ [16], the difference of $4.4$ kcal mole$^{-1}$ corresponding closely to the expected difference in energies due to the heats of adsorption. The two values of the electron affinity are therefore

$$E = E' + D - Q$$

On tungsten $\quad E = 25\cdot1 + 102 - 72 = 55\cdot1 \text{ kcal mole}^{-1}$

On iridium $\quad E = 20\cdot7 + 102 - 66 = 56\cdot7 \text{ kcal mole}^{-1}$

On the other hand $p$-benzoquinone[6] showed an apparent electron affinity (corrected to $0°K$) of $31\cdot2$ kcal mole$^{-1}$ on an iridium filament and of $32\cdot2$ kcal mole$^{-1}$ on a filament of tungsten carbide. These two values are so close that it appears that benzoquinone cannot form a negative ion by a type IV process, and that direct capture is occurring.

### 3.8 Consideration of the Entropy of the Reaction

If the ratio $j_e p / j_i$ represents a true equilibrium constant it would be proper to evaluate not only the energy but also the entropy of ion formation.

$$R \log \left( \frac{j_e}{j_i} \right) + R \log p = \Delta S - \frac{E'}{T} \tag{3.13}$$

where

$$E' = - \frac{R \, \mathrm{d} \log (j_e p / j_i)}{\mathrm{d}(1/T)} \tag{3.14}$$

Hence

$$\Delta S = \frac{E'}{T} + R \log \left( \frac{j_e}{j_i} \right) + R \log p \tag{3.15}$$

If $j_e p / j_i$ represents the ratio of the rate of negative ion formation to that of the emission of electrons, the quantity $\Delta S$ defined by equation (3.15) cannot be regarded as a true entropy, but rather as the ratio of the pre-exponential factors of the rate constants. In the theory of rate processes (Chapter 4) this ratio is regarded as a difference of entropies of activation and leads to an expression for the entropy of negative ion formation which differs from equation (3.15) only in a term in $\log kT$.

Experience has shown that each of the three mechanisms of ion formation gives a different value for $\Delta S$ calculated from equation (3.15). The entropy of ion formation has been calculated for 60 ions whose mechanism of formation is known, so that the value of $\Delta S$ may be used to identify the mechanism in further studies. The results are plotted as a histogram (Figure 3.2); the number of results within a 5 cal deg$^{-1}$ mole$^{-1}$ range are plotted against the mid-point of the entropy range. The values for $\Delta S$ for each mechanism are

Dissociative capture without adsorption (Type III)

$$\Delta S = -18 \cdot 6 \pm 5 \cdot 8 \text{ cal deg}^{-1}$$

Dissociative capture with adsorption (Type IV)

$$\Delta S = +10 \cdot 6 \pm 5 \cdot 9 \text{ cal deg}^{-1}$$

Direct capture (Types I and II)

$$\Delta S = +35 \cdot 5 \pm 12 \cdot 6 \text{ cal deg}^{-1}$$

The numerical values depend upon the units of pressure used in the calculation, which here are mm Hg. The values therefore differ from those of Chapter 4, which are based on c.g.s. units of pressure.

This method of diagnosis is not infallible, as the separation of the ranges is not clear, but it provides a useful check on the other evidence. It does, in effect, set a quantitative measure to the qualitative discussion of the accessible limits of observation given in section 3.6.

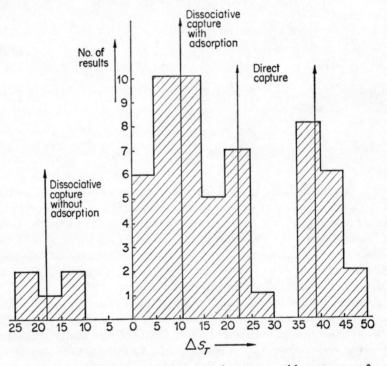

Figure 3.2  Histogram relating reaction type with entropy of reaction

**REFERENCES**

1. Mayer, J. E. and Sutton, P. P., *J. Chem. Phys.*, **3**, 20 (1935).
2. Farragher, A. L., *Ph.D. Thesis*, The University of Aston in Birmingham, 1966.
3. Page, F. M., *Trans. Faraday Soc.*, **57**, 359 (1961).
4. Farragher, A. L., Page, F. M. and Wheeler, R. C., *Disc. Faraday Soc.*, **37**, 203 (1964).
5. Kay, J. and Page, F. M., *Trans. Faraday Soc.*, **60**, 1042 (1964).
6. Farragher, A. L. and Page, F. M., *Trans. Faraday Soc.*, **62**, 2072 (1966).
7. Page, F. M., *Trans. Faraday Soc.*, **57**, 1254 (1961).
8. Page, F. M., *Advan. Chem.*, **36**, 68 (1962).
9. Farragher, A. L., *Unpublished work*.
10. Gaines, A. F., Kay, J. and Page, F. M., *Trans. Faraday Soc.*, **62**, 874 (1966).
11. Gaines, A. F. and Page, F. M., *Trans. Faraday Soc.*, **59**, 1 (1963).
12. Rees, C. W. L., *Dip. Tech. Thesis*, The University of Aston in Birmingham, 1964.
13. Herron, J. T., Rosenstock, H. M. and Shields, W. R., *Nature*, **206**, 611 (1965).
14. Goode, G. C., *Unpublished work*.
15. Cottrell, T. L., *The Strengths of Chemical Bonds*, Butterworths, London, 1958.
16. Burdett, M., *Unpublished work*.

# 4

# The Emission of Electrons and
# Ions as a Rate Process

## 4.1  Introduction

The original interpretation of the action of the magnetron, due to Mayer
and his collaborators[1], postulated a dynamic equilibrium at the surface
of the filament, an equilibrium which was analysed by the application
of statistics. The results obtained when the interpretation was extended
from the halogens to oxygen were not entirely satisfactory and interest
in the method dropped. Page[2], using the alternative, kinetic, formulation
described earlier, succeeded in extending the theory to cases where interaction with the filament occurred, but only by including an empirical rate
constant. Later, Farragher[3] extended the kinetic approach by applying
the theory of rate processes to the emission of ions and electrons, and
developed the use of the entropy of the process as a diagnostic test of the
mechanism of ion production. The fundamental advance made was the
realization that the processes of ion and electron emission were separate
rate processes, each characterized by its own temperature coefficient.

## 4.2  The Theory of Rate Processes

The application of statistical mechanics to the calculation of the rate of
a process, as developed by Glasstone, Laidler and Eyring[4], involves two
steps, the definition of a transition state, and the calculation of the frequency with which molecules pass through this stage. If the reactants
are in a state denoted by the symbol A, in reacting they pass through a
state of maximum energy, the so-called transition state, denoted by B.

Reactants may pass from this state forward, to complete the reaction, or they may revert to the original state A.

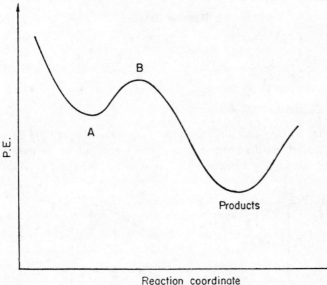

Figure 4.1  The transition state

Provided that the rate of forward passage is small compared with the rate of reversion, the states A and B may be regarded as being in equilibrium so that an equilibrium constant may be set up

$$K = \frac{[B]}{[A]} = \frac{Q_B}{Q_A} \exp \left[ \frac{-(\beta - \alpha)}{kT} \right] \tag{4.1}$$

where $Q_B$, $Q_A$ are the partition functions of states B and A, and $\beta$ and $\alpha$ the corresponding energies.

The concentration of species in state B is therefore given by

$$[B] = [A] \frac{Q_B}{Q_A} \exp \left[ \frac{-(\beta - \alpha)}{kT} \right] \tag{4.2}$$

The partition function $Q_B$ for state B is postulated to contain one term which represents translation leading to completion of the reaction. This term arises from a weak vibration of frequency $\nu$, which contributes $kT/h\nu$ to the partition function so that

$$Q_B = \frac{kT}{h\nu} Q_B^*$$

or
$$[B] = [A] \frac{kT}{h\nu} \frac{Q_B^*}{Q_A} \exp \left[ \frac{-(\beta - \alpha)}{kT} \right] \tag{4.3}$$

The rate of the process will be given by the frequency with which the energy maximum is crossed, identified with the frequency $\nu$, multiplied by the concentration at the state B

$$\text{Rate} = [B]\nu$$

Hence

$$\text{Rate} = [A]\frac{kT}{h}\frac{Q_B{}^*}{Q_A}\exp\left[\frac{-(\beta - \alpha)}{kT}\right] \tag{4.4}$$

### 4.3 The Emission of Electrons

An electron, leaving the surface of the magnetron filament, is subjected to the potential gradient within the magnetron, and to its own image field in the conducting filament. The combination of these two opposed

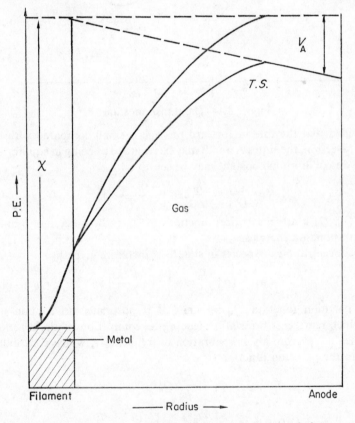

Figure 4.2   The transition state for electron emission

fields leads to a maximum in the potential energy of the electron, an energy barrier known as the Schottky barrier[5] (Figure 4.2).

The field at a point between two concentric cyclinders of radii $r_b$ (potential $= 0$) and $r_a$ (potential $V_a$) is given by

$$E_x = \frac{V_a}{(r_b + x) \log (r_a/r_b)} \tag{4.5}$$

while the image field at the same point is $e/4x$, where $x$ is the distance from the surface of the inner cylinder.

The potential energy of an electron at this point is given by

$$\text{P.E.} = \frac{-e^2}{4x} - \frac{V_a xe}{r_b \log (r_a/r_b)}$$

relative to an electron at an infinite distance from the filament at the potential of the filament.

For the filaments used in the magnetron, $r_b = 1 \cdot 25 \times 10^{-2}$ cm, $r_a = 2 \cdot 25$ cm, $V_a = 120$ V, and the maximum potential energy is $-1 \cdot 65 \times 10^{-2}$ eV occurring at 430 Å from the surface of the filament. No significant error will therefore be introduced if the potential energy at the Schottky barrier is neglected.

An electron at the Schottky barrier satisfies all the requirements for an electron in a transition state for emission from the filament. The system is one of maximum potential energy and crossing the barrier in one direction is an irreversible step leading to completion of the rate process.

The partition function for the electrons in the metal $Q_M$ is given by

$$Q_M = [C_M] \exp \left(\frac{-\delta}{kT}\right) \tag{4.6}$$

where $[C_M]$ is the concentration of electrons, corresponding to A in equation (4.4), and

$$\delta = \frac{3F}{5} + \frac{\pi^2 k^2 T^2}{4F} \tag{4.7}$$

where $F$ is the Fermi energy and given by

$$F = \frac{h^2}{8m_e} \left(\frac{3N}{\pi V}\right)^{\frac{1}{3}} \tag{4.8}$$

Hence, from (4.4), identifying $Q_A$ as $Q_M$

$$\text{Rate} = \frac{kT}{h} Q_B^* \exp \left[\frac{-(\beta - \delta - \alpha)}{kT}\right] \tag{4.9}$$

where $\alpha$ and $\beta$ are the ground state energies of the electron in the metal and in the transition state (taken, as given above, to be negligibly different to the energy at infinity). But

$$\beta - \delta - \alpha = \frac{\chi}{N} \tag{4.10}$$

where $\chi$ is the thermionic electron work function of the filament surface. Hence,

$$\text{Rate} = \frac{kT}{h} Q_{\text{B}}^* \exp\left(\frac{-\chi}{NkT}\right) \tag{4.11}$$

The electron in the transition state will have three degrees of translational freedom. One of these has already been allowed for as the rate determining motion, so that

$$Q_e^* = \frac{2(2\pi m_e kT)}{h^2} \tag{4.12}$$

The factor 2 being included to allow for the spin degeneracy of the electron. Hence,

$$\text{Rate} = 2 \frac{kT}{h} \frac{2\pi m_e kT}{h^2} \exp\left(\frac{-\chi}{NkT}\right) \tag{4.13}$$

or, since the electron current per unit area is the rate of emission of electrons multiplied by the electronic charge

$$j_e = 2e \frac{kT}{h} \frac{2\pi m_e kT}{h^2} \exp\left(\frac{-\chi}{NkT}\right) \tag{4.14}$$

Introducing the transmission coefficient $\bar{d}$, and the gas constant $R = Nk$

$$j_e = \bar{d} \frac{4\pi m_e (kT)^2 e}{h^3} \exp\left(\frac{-\chi}{RT}\right) \cdots \tag{4.15}$$

which is Richardson's equation for thermionic emission, hitherto deduced on thermodynamic or statistical grounds, and not from the theory of rate processes.

### 4.4  The Emission of Ions

It is convenient to divide the problem of calculating the rate of emission of ions from the surface of a filament into two parts, firstly to discuss the rate of emission of ions from a surface fully covered by the precursors of the ions and secondly to discuss the influence of temperature and pressure upon the number of ion precursors present on the surface. It should be stressed that despite the inclusion of a hypothetical fully covered surface in the model, the derivation is intended only to apply to sparsely covered surfaces, where interactions between the adsorbed precursors and desorbing ions may be neglected, as may any interaction between the surface layer and the metal which might modify the energy levels of the metal.

Ions are presumed to be formed by the interaction of electrons in the metal with adsorbed ion precursors, which may of course be either atoms or molecules, resulting in a transition state from which negative ions may be irreversibly desorbed. As in the previous section, the equilibrium constant associated with the formation of the transition state may be written down,

$$K = \frac{[C_B]}{[C_A][C_M]} = \frac{Q_B}{Q_A Q_M} \exp \left[ \frac{-(\beta - \alpha - \sigma)}{kT} \right] \qquad (4.16)$$

where B, A and M refer to the transition state, the adsorbate, and the electron in the metal respectively and $\sigma$ is the ground state energy of the adsorbed ion precursor.

The expression analogous to (4.3) now becomes

$$[C_B] = [C_A][C_M] \frac{kT}{h\nu} \frac{Q_B{}^*}{Q_A Q_M} \exp \left[ \frac{-(\beta - \alpha - \sigma)}{kT} \right] \qquad (4.17)$$

making the same substitution as before [equation (4.6)], and introducing $\bar{d}$, the transmission coefficient, equation (4.11) corresponds to

$$\text{Rate} = \bar{d} \frac{kT}{h} [C_A] \frac{Q_B{}^*}{Q_A} \exp \left[ \frac{-(\beta - \alpha - \sigma - \delta)}{kT} \right] \qquad (4.18)$$

The energy terms have the same meaning as before, except that $\beta$ now represents the system (ion precursor + electron) in the transition state, and not the electron alone.

If, as before, the energy zero at infinity is taken as the energy of the transition state, the algebraic sum of the energy terms may be equated to the sum of the work function $(\chi)$, the heat of desorption $(E_d)$ and the electron affinity $(E)$ of the ion precursor

$$(\beta - \alpha - \sigma - \delta)N = E_d + \chi - E \qquad (4.19)$$

whence

$$\text{Rate} = \bar{d} \frac{kT}{h} Q_B{}^* \frac{[C_A]}{Q_A} \exp \left[ \frac{-(E_d + \chi - E)}{RT} \right] \qquad (4.20)$$

If the process of ion formation represents only a minor perturbation of the adsorption–desorption equilibrium of the ion precursors, the concentration $C_A$ of such precursors may also be evaluated by rate process theory. In the simplest case, when the negative ion is formed by the simple addition of an electron to the precursor, which is itself identical with the gaseous substrate, the rate of adsorption, $U_a$, is given by

$$U_a = [C_s][C_g] \frac{kT_g}{h} \frac{Q_a^*}{Q_s Q_g} \exp\left(\frac{-E_a}{RT_g}\right) \tag{4.21}$$

where $Q_a^*$ is the reduced partition function for adsorption, $C_g$ is the gas phase concentration of the adsorbate, $Q_g$ the partition function for unit volume, $C_s$ the concentration of adsorption sites, and $Q_s$ their partition function per unit area. $E_a$ is the activation energy for adsorption, the temperature $T_g$ is the ambient temperature of the gas, and the partition functions are evaluated at this temperature.

The rate of desorption ($U_d$) will be given by the corresponding equation

$$U_d = [C_A] \frac{kT}{h} \frac{Q_d^*}{Q_A} \exp\left(\frac{-E_d}{RT}\right) \tag{4.22}$$

where $Q_d^*$, $E_d$ refer to the desorption process, and the appropriate temperature $T$ is that of the filament.

At equilibrium, the rates of adsorption and desorption are equal,

$$U_a = U_d$$

$$\frac{[C_A]}{Q_A} = [C_s][C_g] \frac{Q_a^*}{Q_d^*} \frac{1}{Q_s Q_g} \frac{T_g}{T} \exp\left(\frac{-E_a}{RT_g}\right) \exp\left(\frac{E_d}{RT}\right) \tag{4.23}$$

Substituting $[C_s]/Q_s = f(1 - \theta)$, $[C_g] = p/kT_g$,

$$\frac{[C_A]}{Q_A} = f(1 - \theta) \frac{Q_a^*}{Q_d^* Q_g} \frac{p}{kT} \exp\left(\frac{-E_a}{RT_g}\right) \exp\left(\frac{E_d}{RT}\right) \tag{4.24}$$

and the rate of ion emission becomes

$$\text{Rate} = \bar{d}f(1 - \theta) \frac{Q_a^* Q_B^*}{Q_d^* Q_g} \frac{p}{h} \exp\left[\frac{-(\chi - E)}{RT}\right] \exp\left(\frac{-E_a}{RT_g}\right) \tag{4.25}$$

For most adsorption processes at low surface coverage, the activation energy for adsorption ($E_a$) will be zero, and since $\theta$, the fraction of surface covered, is small $f(1 - \theta)$ will approximate to unity. The current will be rate multiplied by charge, so that the ion current density is given by

$$j_i = \frac{\bar{d}pe}{h} \frac{Q_a^* Q_B^*}{Q_d^* Q_g} \exp\left[\frac{-(\chi - E)}{RT}\right] \tag{4.26}$$

where $e$ is the electronic charge.

Separating the translational partition function from the internal partition functions, and assuming that the latter are identical for the gas phase molecule and for the transition states for adsorption and desorption

$$\frac{Q_a{}^*}{Q_g} = \frac{h}{(2\pi M k T_g)^{\frac{1}{2}}}$$

$$\frac{Q_B{}^*}{Q_a{}^*} = \frac{Q_i}{Q_m}$$

where $Q_i$ and $Q_m$ are the internal partition functions of the transition state ion and gas phase molecule respectively, and $M$ is the mass of the ion. Hence

$$j_i = \frac{\bar{d}pe}{(2\pi M k T_g)^{\frac{1}{2}}} \frac{Q_i}{Q_m} \exp\left[\frac{-(\chi - E)}{RT}\right] \tag{4.27}$$

## 4.5   Application to the Magnetron

In work with the magnetron, it is usual to consider the ratio of the electron ($j_e$) to ion ($j_i$) currents, and this ratio may be calculated directly from equations (4.15) and (4.27)

$$\frac{j_e}{j_i} = \frac{4\pi m_e (kT)^2 (2\pi M k T_g)^{\frac{1}{2}}}{h^3 p} \frac{Q_m}{Q_i} \exp\left(\frac{-E}{RT}\right) \tag{4.28}$$

if the transmission coefficients may be assumed to cancel.

Since the apparent electron affinity at the temperature of experiment ($E_T{}'$) is evaluated from

$$E_T{}' = -R\left[\frac{\mathrm{d} \log (j_e/j_i)}{\mathrm{d}(1/T)}\right]_p \tag{4.29}$$

the differentiation of equation (4.28) with respect to $(1/T)$ yields

$$E_T{}' = E + 2RT \tag{4.30}$$

if the ratio $Q_m/Q_i$ is sensibly independent of temperature. This point is developed in detail in section 6.3.

If equations (4.14) and (4.12) are combined the rate of emission of electrons may be written as

$$j_e = \bar{d}e \frac{kT}{h} Q_e{}^* \exp\left(\frac{-\chi}{RT}\right) \tag{4.31}$$

3+

which may be combined with (4.25), if the transmission coefficients are assumed to cancel and $f(1 - \theta)$ to approximate to unity, to give

$$\frac{j_e}{j_i} = \frac{kT}{p} \frac{Q_e{}^* Q_d{}^* Q_g}{Q_a{}^* Q_B{}^*} \exp\left(\frac{E_a}{RT_g}\right) \exp\left(\frac{-E}{RT}\right) \tag{4.32}$$

$$= \left(\frac{kT}{p}\right)\left[\frac{Q_g}{Q_a{}^*} \exp\left(\frac{E_a}{RT_g}\right)\right]\left[\frac{Q_e{}^* Q_d{}^*}{Q_B{}^*} \exp\left(\frac{-E}{RT}\right)\right]$$

$$= \frac{kT}{p} K_1 K_2 \tag{4.33}$$

where $K_1$ and $K_2$ are the concentration equilibrium constants for the reactions

$$A_{\text{ads}}^* \rightleftharpoons A(g)$$
$$A^{-*} \rightleftharpoons A_{\text{des}}^* + e^{-*}$$

This is the equation referred to in section 3.8 for the entropy of ion formation but whereas the development in that section was purely in terms of experimental data, the statistical expressions for the partition functions may be used as a more absolute test of the data.

Equation (4.32) leads, when combined with (3.14), to the relation

$$\Delta S_K = R \log\left[\frac{Q_g Q_e{}^* Q_d{}^*}{Q_a{}^* Q_B{}^*} \exp\left(\frac{E_a}{RT_g}\right)\right] + nR \tag{4.34}$$

where $n$ is defined from

$$E_T' = E + nRT$$

Inserting the values of the fundamental constants, and assuming that for direct capture reactions there is no contribution from the internal degrees of freedom (i.e. $Q_m = Q_i$),

$$\Delta S_K = 4{\cdot}57[18{\cdot}25 + \log_{10} T + \tfrac{1}{2} \log_{10} M] + 3{\cdot}96 \tag{4.35}$$

where $M$ is in atomic mass units and $\Delta S$ in cals deg$^{-1}$ mole$^{-1}$.

This value of $\Delta S_K$ represents the contribution to $\Delta S_T$, the entropy calculated from equation (3.14) of the simple kinetic terms, and the values of $\Delta S_T$ and $\Delta S_K$ are set out in Table 4.1 together with $\log_{10} W$, which represents the internal entropy contributions, and is here defined by

$$\Delta S_T = \Delta S_K - \log_e W$$

The loss of a rotational degree of freedom would introduce a term $0{\cdot}30 + \tfrac{1}{2} \log_{10} (I_x)$ within the bracket of equation (4.35) which would be equivalent to a factor of $10^{-2}$ in $W$. The loss of two rotational degrees of freedom therefore appears to be very common.

Table 4.1

| Substrate | $\Delta S_T$ | $\Delta S_K$ | $\log_{10} W$ |
|---|---|---|---|
| $C_6F_6$ | 96·2 | 99 | +0·57 |
| $SF_6$ | 100·9 | 99 | −0·53 |
| NO | 91·2 | 97 | +1·33 |
| *p*-benzoquinone | 91·1 | 100 | +1·90 |
| fumaronitrile | 88·3 | 98 | +1·49 |
| *o*-phthalonitrile | 95·5 | 99 | +0·72 |
| $CCl_4$ | 114·8 | 97·5 | −3·76 |
| $CHCl_3$ | 106·2 | 98 | −1·73 |
| Tetracyanobenzene | 109 | 98 | −2·33 |
| Tetracyanopyridine | 113·1 | 98 | −3·23 |
| Tetracyanoethylene | 112 | 97 | −3·41 |
| Chloranil | 106·9 | 98 | −1·73 |
| Tetracyanoquinodimethane | 118 | 99 | −3·94 |
| Hexacyanobenzene | 121 | 100 | −4·60 |

The errors inherent in the experimental determination of the entropy, notably in the measurement of the pressure, make pointless any attempt to determine electron affinities by applying these statistical arguments, and any attempts to refine the calculations of $W$ are also of doubtful value. The agreements obtained on the zero order approximation are, however, so striking as to give confidence in the accuracy of the theory as thus developed, and in the direct results as applied in Chapters 3 and 6.

**REFERENCES**

1. Mayer, J. E. and Sutton, P. P., *J. Chem. Phys.*, **3**, 20 (1935).
2. Page, F. M., *Trans. Faraday Soc.*, **56**, 1742 (1960).
3. Farragher, A. L., *Ph. D. Thesis*, The University of Aston in Birmingham, 1966.
4. Glasstone, S., Laidler, K. J. and Eyring, H., *The Theory of Rate Processes*, McGraw-Hill, New York, 1941.
5. Schottky, W., *Ann. Phys.* **44**, 1011 (1914).

# The Effects of Adsorption

## 5.1 Measurements at Real Pressures

The emission of both electrons and negative ions has been assumed to be proceeding without interference from the vapour within the magnetron. This assumption is valid only at low pressures, since negative ions may be formed by attachment in the gas phase and also the motion of ions and electrons may be impeded by collisions. The chance that attachment, direct or dissociative, will occur at a collision has been shown for many gases to be of the order of $10^{-4}$ [1,2] so that negative ion formation, other than at the filament surface, may be disregarded unless collisions are frequent. The whole analysis of the previous chapter is based on the presumption that such collisions are rare and the apparatus is so designed that the mean free path of an ion is comparable with the dimensions of the apparatus at a pressure of $5 \times 10^{-3}$ mm Hg while that of the electron will be greater. The chance that one collision will occur is therefore small, and the assumptions made in the previous chapters are valid.

Nevertheless, strong effects of the vapour on both the electron and ion currents are frequently observed, and these are attributed to adsorption of the vapour on the surface of the filament. The effects may conveniently be considered under three headings: gases which increase the work function of the surface, gases which decrease it and the effects of adsorption on the ion current.

## 5.2 Elevated Work Functions

It has been known from the earliest work upon the thermionic emission of electrons that the presence of traces of adsorbable gases, particularly oxygen, would raise the work function of a metal surface. Many examples of abnormally high work functions are to be found in the literature usually being associated with high pre-exponential factors in Richardson's equation[3]; it is noteworthy that these very high work functions are found only when measurements are made by thermionic methods. The direct measurement of the surface potential of a surface upon which the gas is adsorbed, relative to that of the clean surface, indicates a much smaller increase in work function. It would be easy to dismiss the early work as unreliable and the phenomena as irreproducible, but certain measurements made by Kingdon[4] in 1924 have been reproduced in these laboratories forty years later. It is apparent therefore that these abnormally high work functions and pre-exponential terms are real and reproducible phenomena which are susceptible to an analytic explanation.

The basic theory of the emission of electrons from a hot metal surface was advanced by Richardson[5] using thermodynamic arguments and was restated by Fowler[6] on a quantum mechanical basis. The two derivations differ only in a factor of 2 arising from the electron spin, the current density in amp cm$^{-2}$ being given by

$$j_e = \frac{\bar{d}4\pi m_e(kT)^2 e}{h^3} \exp\left(\frac{-\chi}{RT}\right) \tag{5.1}$$

where $\bar{d}$ is the transmission coefficient, $m_e$ the electron mass, $k$ Boltzmann's constant, $T$ the temperature of the metal in °K, $e$ the electronic charge, $h$ is Planck's constant, $R$ the gas constant and $\chi$ the electron work function. This equation was derived in the previous chapter on a purely kinetic basis.

The majority of surfaces upon which thermionic measurements are made are polycrystalline, having different faces with different work functions which contribute unequally to the total thermionic current. A polycrystalline surface of area $A$ cm$^2$ may be considered to be composed of $i$ patches, the $i$th patch having an area $\theta_i A$ cm$^2$ and work function $\chi_i$. If the emission current from each patch ($j_i$) is given by the appropriate Richardson equation $j_T$ the total current may be written as

$$j_T = \sum_i j_i = \bar{d}_m BAT^2 \sum_i \theta_i \exp\left(\frac{-\chi_i}{RT}\right) \tag{5.2}$$

where $j_T$ is the total current emanating from $A$ cm$^2$ of surface, $\bar{d}_m$ is the mean transmission coefficient and $B = 4\pi m_e k^2 e/h^3$. The experimental thermionic work function $\chi'$ is defined by

$$\chi' = \frac{-R \, \mathrm{d} \ln (j_T/T^2)}{\mathrm{d}(1/T)} \tag{5.3}$$

whence by differentiation of equation (5.2), assuming that the transmission coefficient is independent of temperature,

$$\chi' = \sum_i \phi_i \chi_i + \frac{RT^2 \sum_i \phi_i \, \mathrm{d} \ln \theta_i}{\mathrm{d}T} - \frac{T \sum_i \phi_i \, \mathrm{d}\chi_i}{\mathrm{d}T} \tag{5.4}$$

where $\phi_i = j_i/j_T$. For clean surfaces $\chi_i$ and $\theta_i$ may be considered to be temperature independent so that

$$\chi' = \sum_i \phi_i \chi_i \tag{5.5}$$

The thermionic work function $(\chi')$ is not the same as the mean work function derived from contact potential measurements $(\bar{\chi})$ which is defined as

$$\bar{\chi} = \sum_i \theta_i \chi_i \tag{5.6}$$

since the weighting is according to emission in $\chi'$, but according to area in $\bar{\chi}$.

It is not possible to carry the analysis of the surface into patches of differing work function to the limit of atomic dimensions. Under experimental conditions of small (40 V cm$^{-1}$) applied fields the Schottky barrier lies about 400 Å from the surface, and electrons between the surface and barrier must be regarded as being in equilibrium with the underlying surface. Only patches whose dimensions are significantly larger than the distance of the Schottky barrier may properly be included in the summation. If the patches are smaller than this the assumption of a discontinuous distribution fails since their separate influences on electrons at the Schottky barrier will not be resolved, and the whole area must be treated as one patch with a mean work function.

A gas which may be strongly adsorbed will change the work function of each patch, and since the amount adsorbed will be a function of temperature both $\chi_i$ and $\phi_i$ will vary. In the limiting case of a two-patch surface, one covered to an extent $\theta$ by an adsorbed gas which raises the work function sufficiently for the emission current from the surface to be neglected, and one uncovered

$$\chi' = \chi_0 + \frac{RT^2 \, \mathrm{d} \ln (1 - \theta)}{\mathrm{d}T} \tag{5.7}$$

which, utilizing the Langmuir relation for a chemisorbed monolayer[7]

$$\frac{(1 - \theta)}{\theta} = \frac{k}{p^x} \exp\left(\frac{-q}{RT}\right) \tag{5.8}$$

where $q$ is the heat of adsorption, gives

$$\chi' = \chi_0 + \theta q \tag{5.9}$$

The assumption of a two-patch surface is, however, an unrealistic approach as it assumes a discontinuous change in work function in the presence of the adsorbed material and the treatment also ignores the variation of $q$ with $\theta$ and hence $T$ which is certain to occur at high surface coverages.

If Richardson's equation for the covered surface is written in the form

$$j_T = \bar{d}_m BAT^2 \exp\left(\frac{-\chi_m}{RT}\right) \tag{5.10}$$

$$\chi' = \frac{\chi_m - T\,\mathrm{d}\chi_m}{\mathrm{d}T} \tag{5.11}$$

$\chi_m$ may be written as $\chi_m = \chi_0 + \Delta\chi_m$, where $\chi_0$ is the experimental thermionic work function of the surface in the absence of any gas. Mignolet[8] has shown that, if the change in the differential heat of adsorption may be attributed entirely to the reduction in the energy of the highest occupied level of the adsorbate, then $\Delta\chi = -\Delta q$, where the differential heat of adsorption of the gas is $q = q_0 - \Delta q$ and $\Delta q = 0$ when $\theta = 0$. Hence combining equations (5.8) and (5.10)

$$j_T = \bar{d}_m \frac{BAT^2 k}{p^x} \frac{\theta}{1 - \theta} \exp\left[\frac{-(\chi_0 + q_0)}{RT}\right] \tag{5.12}$$

For non-dissociative adsorption $x = 1$ and for dissociative adsorption and associative desorption $x = 0.5$.

If the term $\theta/(1 - \theta)$ is a slowly varying function of temperature, differentiation of equation (5.12) gives

$$\chi' = \chi_0 + q_0 \tag{5.13}$$

However, if equation (5.10) is assumed and the work function calculated from the measured values of $j_T$, $A$ and $T$

$$\chi_m = \chi_0 + \Delta\chi_m \tag{5.14}$$

equations (5.13) and (5.14) define the results of thermodynamic and statistical methods of measuring the thermionic work function respectively. Since $q_0 > \Delta\chi$ the second law estimate will be the greater, as was noted previously.

It may be argued that the term $\theta/(1 - \theta)$ in equation (5.12) cannot be a slowly varying function of temperature since the dependence is given by equation (5.8). The relation put forward by Mignolet[8] ($\Delta\chi_m = q$) indicates that the total (work function + heat of adsorption) is a constant, equal to the value at zero coverage

$$\chi_m + q = \chi_0 + q_0 \tag{5.15}$$

Equation (5.9) (with $\theta = 1$) will therefore be valid, despite any variation in $\chi$ and $q$ with surface coverage, since these variations will cancel. This, in turn, implies that $\theta/(1 - \theta)$ is indeed a slowly varying function of temperature, having a value of approximately unity, and that $\theta$ is of the order of 0·5. Riemann[9] has argued that this is in fact usually the case and that high degrees of surface coverage are rarely achieved.

### 5.3  Experimental Observations

A number of gases have been examined in sufficient detail[10] to merit the application of the theoretical analysis developed in the previous section. The results are summarized in Table 5.1.

A comparison of equation (5.13) in the presence and absence of gas shows that one may write the ratio of the currents at a fixed temperature as

$$\frac{j_T}{j_0} = \frac{k}{p^x} \frac{\theta}{1 - \theta} \exp\left(\frac{-q_0}{RT}\right) \tag{5.16}$$

where $k$ is the pre-exponential factor, and $x$ the order of pressure dependence derived from the Langmuir equation. The general transition state theory of non-dissociative adsorption indicates that $x$ should be unity, $k$ of the order of $10^{11}$–$10^{15}$, and $q_0$ is then the molecular heat of adsorption, whereas the theory for dissociative adsorption indicates values of $x$ of 0·5, $k$ of $10^3$–$10^7$ with $q_0$ being half the heat of adsorption. Utilizing equation (5.16) and the above premises, molecular heats of adsorption can be determined by the magnetron technique; the results are included in Table 5.1.

The only gas where a direct comparison between magnetron values and calorimetric values can be made is oxygen. Oxygen on tungsten shows a square-root pressure dependence in the magnetron and gives a value for the molecular heat of adsorption of 192 kcal mole$^{-1}$ as compared with the calorimetric value of 194 kcal mole$^{-1}$. All other published data involve unjustified assumptions, but yield values which correspond more or less closely to the values in Table 5.1.

Table 5.1

| Gas | NO₂ | NO | | CO | | O₂ | | O₂ |
|---|---|---|---|---|---|---|---|---|
| Filament | Tungsten | Tungsten | | Tungsten | | Tungsten | | Iridium |
| $\chi$ kcal mole$^{-1}$ | 224 | 153 | 223 | 157 | 222 | 202 | 200 | 154 |
| $T$ °K | 1792 | 1790 | 1640 | 1750 | 1610 | 1750 | | 1590 |
| $p$ mm Hg | $3 \times 10^{-3}$ | $1.8 \times 10^{-3}$ | | $8 \times 10^{-3}$ | | $1.1 \times 10^{-3}$ | $5 \times 10^{-5}$ | $1.2 \times 10^{-3}$ |
| $\chi_0$ kcal mole$^{-1}$ | 103 | 91 | | 104 | | 105 | | 81 |
| $q_0$ kcal mole$^{-1}$ | 121 | 62 | 132 | 53 | 118 | 96 | | 73 |
| $x$ | 0.5 | 0.5 | 1.0 | 0.5 | 1.0 | 0.5 | | 0.5 |
| $\log_{10} k$ | 10.0 | 5.3 | 14.4 | 4.6 | 13.6 | 8.9 | | 6.4 |

3*

## 5.4   Lowered Work Functions

The previous sections dealt with those cases where the adsorption of a gas raised the work function of a surface. The model used was that of a patch surface, the temperature coefficient of electron emission being made up of two terms, one arising from the normal increase in emission with increasing temperature, and the other from the increase in emitting surface with increasing temperature. A number of systems have been observed in which adsorption lowers the work function of the surface. One of the most notable of these is the molecule tetracyanoethylene, which on tantalum carbide is reversibly adsorbed, and lowers the work function of the surface by about one volt. Some measurements on this system are shown in Figure 5.1.

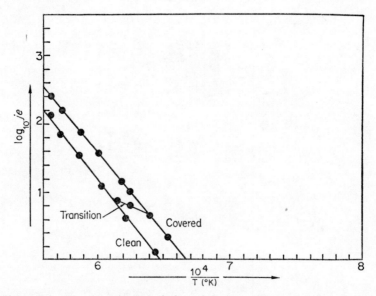

Figure 5.1   The thermionic emission of electrons from a tantalum carbide surface in the presence of tetracyanoethylene

The transition between the clean and covered surface lines is clearly demonstrated, and points could be obtained in the intermediate range. The phenomenon was completely reversible and, despite the complexity of the molecule concerned, there was no trace of decomposition, the negative ions produced being solely of mass 128[11].

The treatment given in section 5.2 is equally applicable to the case where the work function is lowered, but this case may more profitably be examined from a different point of view. The transmission coefficient $\bar{d}_m$ in equation (5.3) was not considered in any detail, since it is essentially less than unity and the apparent pre-exponential factors were all increased manyfold over the corresponding clean surface factors. When the work function is lowered, however, there is an accompanying reduction in the pre-exponential factor, and one contributory cause may be the transmission coefficient.

## 5.5 The Transmission Coefficient of an Ideal Double Layer

The detailed theoretical treatment of the rate of emission of electrons from surfaces with various types of energy boundaries has been given by Fowler[6] on which the following discussion is based.

If the number of electrons with energies between $\zeta$ and $\zeta + d\zeta$ incident on a unit area of a surface in unit time is $N(\zeta)\,d\zeta$ (considering only the component normal to the surface), and $D(\zeta)$ is the chance that such an electron will emerge, then the rate of emission of electrons per unit area is $j_e$ where

$$j_e = \int_{\chi_0}^{\infty} N(\zeta)D(\zeta)\,d\zeta \tag{5.17}$$

where $\chi_0$ is the work function of this surface.

The mean transmission coefficient $(\bar{d})$ is defined by writing this integral as

$$\bar{d}\int_{\chi_0}^{\infty} N(\zeta)\,d\zeta = \int_{\chi_0}^{\infty} N(\zeta)D(\zeta)\,d(\zeta) \tag{5.18}$$

so that

$$\bar{d} = \frac{\int_{\chi_0}^{\infty} N(\zeta)D(\zeta)\,d(\zeta)}{\int_{\chi_0}^{\infty} N(\zeta)\,d\zeta} \tag{5.19}$$

The calculation of the chance that electron may emerge is a problem of some complexity and no complete solution has been given, but some results are given in Fowler[6]. Two cases of particular interest to the present considerations arise when the ideal abrupt boundary (Figure 5.2a) is modified by barriers which reduce the work function $\chi_0$ by an amount $\Delta\chi$. The calculation for the square boundary (Figure 5.2b) has been made exactly, and leads to a transmission coefficient given by

$$\bar{d} = 8\left[\pi k T\left(1 - \frac{\Delta\chi}{\chi_0}\right)\right]^{\frac{1}{2}} \exp\left[-2\kappa l(\Delta\chi)^{\frac{1}{2}}\right] \tag{5.20}$$

where $\kappa^2 = 8\pi^2 m/h^2$ and $l$ the thickness of the barrier.

The value of the transmission coefficient for the ideal double layer (Figure 5.2c) will be of the same form, except that the coefficient 2 in the exponential becomes 4/3.

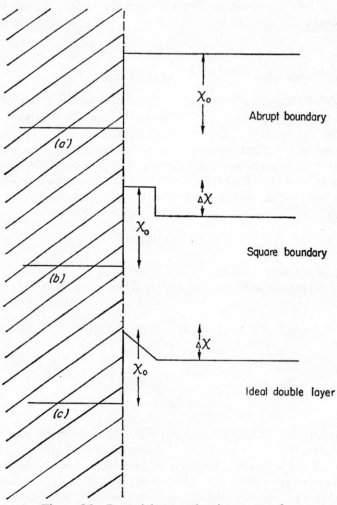

Figure 5.2   Potential energy barriers at a surface

It is most improbable that any of these sharply discontinuous barriers can be a good representation of the real surface field, irrespective of the superposition of the image field, but it can be shown generally that, whatever the detailed shape of the barrier, the same formal type of expression for the mean transmission coefficient is obtained and since the expression is dominated by the exponential term, which is but little affected by the barrier shape, the expression for the ideal double layer will be used here.

## 5.6 Work Functions in the Presence of Tetracyanoethylene[10]

If equation (5.2) is used to compare the electron currents with and without the presence of an adsorbed film, and considering only two types of patch, clean and covered, one may write

$$\frac{j_T}{j_0} = \frac{\bar{d}_i \theta_i \exp\left(-\chi_i/RT\right)}{\bar{d}_0 \exp\left(-\chi_0/RT\right)} + (1 - \theta_i) \tag{5.21}$$

Denoting the ratio of the transmission coefficients by $\alpha$ and the lowering of the work function $(\chi_0 - \chi_1)$ by $\Delta\chi$

$$\frac{j_T}{j_0} = \alpha\theta_i \exp\left(\frac{\Delta\chi}{RT}\right) + (1 - \theta_i) \tag{5.22}$$

Noting that $j_T/j_0$ is large compared to unity

$$RT \log\left(\frac{j_T}{j_0}\right) = RT \log\left(\alpha\theta_i\right) + \Delta\chi \tag{5.23}$$

Figure 5.3, based on the data of Figure 5.1 and other material, suggests that, for tetracyanoethylene on tantalum $\Delta\chi = 22 \cdot 7$ kcal and $\alpha\theta$ has a value of about $3 \cdot 3 \times 10^{-2}$. A set of data referring to the adsorption of tetracyanoethylene on various metals derived in this manner is given in Table 5.2.

When this data is compared with the theoretical expression for the transmission coefficient, an estimate has to be made of the thickness of the barrier. If this is calculated from the value of $\alpha\theta$ obtained the thickness is in all cases unreasonably large, except for the specially conditioned filaments of tungsten carbide which had been exposed to the vapour of fumaronitrile before use. If the filament so conditioned was heated until desorption occurred, and then regenerated without fumaronitrile, the behaviour was different, as shown. This special conditioning of the surface was observed, to a less dramatic extent, with most filaments, and has been referred to as 'ageing', 'conditioning' or 'surface recrystallization'; it is a prerequisite for the formation of a stable and reproducible surface on which to study adsorption.

Alternatively, the estimate of the barrier thickness may be based on the model of the double layer formed by the adsorbed molecule and the surface layer of atoms. This distance, if estimated from the covalent radii, will be a slight underestimate, and since the value thus calculated, $(4 \cdot 5 \text{ Å})$, is

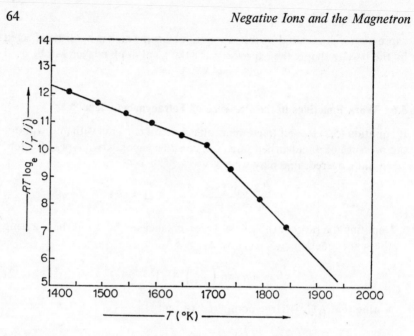

Figure 5.3   Relative thermionic electron currents in the system
TCNE/TaC

Table 5.2   The effect of adsorption of tetracyanoethylene on electron
emission

| Filament | Pressure (mm Hg) | $\Delta\chi$ (kcal) | $\alpha\theta$ | $l$ (Å) | $\theta$ |
|---|---|---|---|---|---|
| Tantalum | $1\cdot0 \times 10^{-5}$ | 11·7 | 0·0931 | 7·6 | 0·32 |
| Tantalum | $2\cdot0 \times 10^{-5}$ | 19·6 | 0·0386 | 7·3 | 0·28 |
| Tantalum | $1\cdot8 \times 10^{-4}$ | 22·0 | 0·0334 | 7·0 | 0·32 |
| Tantalum | $5\cdot0 \times 10^{-4}$ | 22·0 | 0·0330 | 7·0 | 0·32 |
| Tantalum | $1\cdot0 \times 10^{-3}$ | 21·7 | 0·0334 | 7·0 | 0·32 |
| Tungsten Carbide | $1\cdot5 \times 10^{-4}$ | 6·8 | 0·346 | 6·5 | 0·64 |
| Tungsten Carbide* | $1\cdot0 \times 10^{-4}$ | 22·7 | 0·108 | 5·2 | 1·0 |
| Tungsten Carbide* | $8\cdot0 \times 10^{-5}$ | 22·5 | 0·101 | 5·3 | 1·0 |
| Tungsten Carbide* | $2\cdot0 \times 10^{-5}$ | 22·9 | 0·108 | 5·2 | 1·0 |
| Molybdenum Carbide | $1\cdot1 \times 10^{-4}$ | 6·3 | 0·329 | 6·9 | 0·56 |
| Molybdenum Carbide | $1\cdot1 \times 10^{-4}$ | 30·0 | 0·0078 | 7·8 | 0·14 |
| Tantalum Carbide | $2\cdot0 \times 10^{-4}$ | 14·0 | 0·0658 | 7·6 | 0·29 |
| Rhodium⎫ Iridium ⎬ Platinum⎭ | No effects observed | | | | |

The three filaments marked * had been specially conditioned before use.

close to the minimum value of $l$ which was obtained from $\alpha\theta$ (5·3 Å) the latter value may be accepted. This is equivalent to saying that on these surfaces there is a complete close packed coverage. With an accepted value of $l$ the corresponding value of $\alpha$ may be calculated and the fraction of surface covered ($\theta$) evaluated; the fractional values obtained are completely reasonable. A molecule as large as tetracyanoethylene, occupying 30 sites (6 × 5) must be surrounded by a clear space, if an adjacent molecule is separated by slightly less than the molecular dimensions, a third molecule cannot be placed between. A naive model may be constructed in which every molecule is surrounded by a band of width equal to half the molecular diameter, so that the maximum value of $\theta$ would be 0·25. Only if the film were highly mobile, and the conditions very favourable, could a close packed state be achieved.

## 5.7 Adsorption and Ion Formation

The phenomena discussed in the previous sections are concerned with the covering of a substantial part of the surface with an adsorbed monolayer. Since experiments are conducted at temperatures of about 1500°K, and pressures of $10^{-4}$ mm Hg, the heats of adsorption needed to produce significant surface coverages are of the order of 100 kcal mole$^{-1}$; the phenomena are thus restricted to conditions of strong atomic chemisorption, or to the adsorption of highly polar molecules. It was noticed, however, at an early stage in the development of the technique, that the temperature coefficients of the ion currents from certain classes of molecules, notably those containing hydrogen, appeared to include a contribution from the heats of adsorption of fragments of the molecules on the filament. This was shown to bring the observed and predicted temperature dependences for HCl, where all energies are known, into coincidence; and has also been demonstrated in many other cases, discussed below. If the actual process of ion formation involves the transfer of a part of the molecule from the ion precursor to the surface, the energies binding this fragment both to the ion precursor and to the surface are necessarily involved, but if the fragment is an equilibrium state in relation to the surface, adsorption will play no part in determining the overall energetics of the ion formation. If, however, the process of desorption is 'insulated' from the process of ion formation, for example by the interposition of a slow kinetic step such as diffusion across the surface, then the energetics of adsorption will contribute to the total. Such a slow step will be most important when the surface coverage is low, so that the adsorbed fragment has far to go before encountering a partner, and also for very simple fragments, particularly atoms, which cannot desorb without a partner.

### 5.8   Adsorption of Hydrogen Atoms

The first direct evidence for the adsorption of a hydrogen atom occurring during the process of ion formation was obtained during an examination of the halogen acids, HCl and HBr[12], on a tungsten surface. In the case of HCl, negative ions may be formed by several processes.

$$HCl + e \rightarrow HCl^- \qquad E' = E_{HCl}$$
$$\tfrac{1}{2}H_2 + Cl^- \qquad E' = E_{Cl} - D_{H-Cl} + \tfrac{1}{2}D_{H-H}$$
$$\tfrac{1}{2}Cl_2 + H^- \qquad E' = E_H - D_{H-Cl} + \tfrac{1}{2}D_{Cl-Cl}$$
$$H + Cl^- \qquad E' = E_{Cl} - D_{H-Cl}$$
$$Cl + H^- \qquad E' = E_H - D_{H-Cl}$$
$$H_{ads} + Cl^- \qquad E' = E_{Cl} + Q_H - D_{H-Cl}$$

Except for the first process (which was discussed in the introduction and shown to have a negative electron affinity) and the last, the change in enthalpy $E'$ of each process can be calculated exactly, and all differ widely from the observed value for $E'$ of 61 kcal mole$^{-1}$ at 1810°K. A value for the heat of adsorption of the hydrogen atom on a tungsten surface of 73·5 kcal mole$^{-1}$ was calculated by Eley[13], from the molecular heat of adsorption of hydrogen, and this value, when used in conjunction with the known bond energy in HCl (102·2 kcal mole$^{-1}$ [21]) and electron affinity of Cl (83·2 kcal mole$^{-1}$ [20]), predicts an overall value for $E'$ of 54·5 kcal mole, in reasonable agreement with that observed. A similar agreement was found when HBr was considered. There are relatively few other substances for which all the required data are available, but after considering the overall energetics of ion formation in water, hydrogen chloride, hydrogen bromide and ammonia, the mean value for the heat of adsorption of a hydrogen atom on tungsten was established as 73·0 ± 1·1 kcal mole$^{-1}$ at 0°K, or 72·0 kcal mole$^{-1}$ at a mean filament temperature of 1500°K[14] in close agreement with the value calculated by Eley.

Table 5.3   Heats of adsorption of the hydrogen
atom at $T = 1500$°K

| Filament Material | Heat of Adsorption |
|---|---|
| Tungsten | 72 kcal mole$^{-1}$ |
| Tungsten Carbide | 72 kcal mole$^{-1}$ |
| Tantalum | 71 kcal mole$^{-1}$ |
| Iridium | 66 kcal mole$^{-1}$ |
| Platinum | 69 kcal mole$^{-1}$ |

The heats of adsorption of a hydrogen atom on some commonly used filament materials are collected in Table 5.3.

## 5.9 Adsorption of Halogens

It was, for a long time, thought that only the hydrogen atom could participate in the process of ion formation through its adsorption on the filament surface, but evidence has gradually accumulated that other atoms, and even radicals[15,16], can contribute their heat of adsorption to the overall energetics.

During the study of ion formation in chloroform[6] it was observed that one process by which ions could be formed was

$$CHCl_3 + e \rightarrow CCl_3^- + H_{ads} \qquad \text{where } E' = E_{CCl_3} - D_{CCl_3-H} + Q_H$$

This process, which could be identified by having slightly different values of $E'$ on different filaments, gave an unequivocal value for the electron affinity of the $CCl_3$ radical.

When carbon tetrachloride was studied, an analogous process

$$CCl_4 + e \rightarrow CCl_3^- + Cl_{ads} \qquad \text{where } E' = E_{CCl_3} - D_{CCl_3-Cl} + Q_{Cl}$$

appeared to take place, and from the overall energetics of this process, the heat of adsorption of a chlorine atom on platinum was deduced to be $35 \cdot 3$ kcal mole$^{-1}$ at $T = 0°K$[17].

The heat of adsorption of the bromine atom on platinum was also found by comparing the corresponding processes in bromoform and carbon tetrabromide[18], and though the inferences are less clear cut, the heat of adsorption appears to be close to 25 kcal mole$^{-1}$. Some studies of ions formed on iridium filaments suggest that the heat of adsorption is 17 kcal mole$^{-1}$, but there is some uncertainty about the bond energies to be used in the calculations.

The heat of adsorption of fluorine on platinum has similarly been found by a comparison of processes in fluoroform and carbon tetrafluoride. This study is reinforced by observations on other derivatives of carbon tetrafluoride[19].

The apparent electron affinities and mean temperatures observed were

$$CF_4/Pt \qquad E' = -18 \cdot 2 \pm 1 \cdot 1 \text{ kcal mole}^{-1} \ \bar{T} = 1390°K \quad (1)$$

$$CF_3H/Pt \quad E' = 16 \cdot 7 \pm 1 \cdot 8 \text{ kcal mole}^{-1} \ \bar{T} = 1342°K \quad (2)$$

The mechanisms postulated for these processes were

$$(1) \quad CF_4 + e \ \rightarrow CF_3^- + F_{ad} \quad E' = E - D + Q_F$$

$$(2) \quad CF_3H + e \rightarrow CF_3^- + H_{ad} \quad E' = E - D + Q_H$$

Correcting $E'$ to $0°K$ and substituting $Q_H = 68$ gives

$$Q_{F/Pt} = 50 \cdot 2 \pm 2 \cdot 9 \text{ kcal mole}^{-1} \text{ at } 0°K$$

There are very few molecular heats of adsorption which can be used to provide comparable atomic heats of adsorption. Gaines, Kay and Page[17] have pointed out that if the heats of adsorption of hydrogen and oxygen atoms on platinum are plotted against those on carbon, the atomic heat of adsorption of chlorine on carbon, calculated from the observed molecular heat of adsorption, when plotted on the straight line drawn through the two points for hydrogen and oxygen, indicates an atomic heat of adsorption of chlorine on platinum in reasonable agreement with that observed (Figure 5.4).

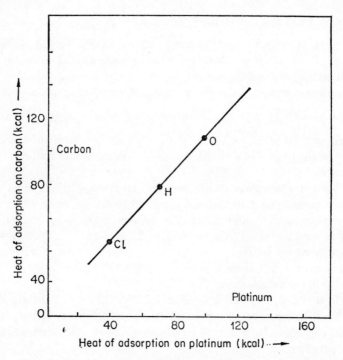

Figure 5.4    Correlation between the atomic heats of adsorption on differing surfaces

## 5.10   Heats of Adsorption of Radicals

Since the adsorption of atoms can provide a significant contribution to the observed electron affinity, it might at first be thought that many other molecular fragments could be adsorbed on the filaments and make similar contributions. There is, however, little evidence that this type of behaviour occurs.

Kay and Page, in discussing the electron affinity of the hydroxyl radical[15], claimed that one process by which ions were formed in hydrogen peroxide vapour appeared to involve the adsorption of a hydroxyl radical on the filament.

They derived the heat of adsorption of hydroxyl on platinum from the increased work function of the filament as 68 kcal mole$^{-1}$, and obtained values for the electron affinity which agreed well with those obtained from a non-adsorptive process occurring in a different temperature range.

The only other radical for which any adsorptive process has been suggested was the methyl radical. A survey of four derivatives of methyl (nitromethane, methanol, methylamine and acetonitrile) by Ritchie and Wheeler[16] gave results which, when the known electron affinities of the radicals $NO_2$, OH, $NH_2$ and CN were inserted, were consistent with a heat of adsorption of the methyl radical on tungsten of 48 kcal mole$^{-1}$. No direct evidence to support this figure, analogous to the molecular heat of adsorption of hydrogen, is available, but some indirect evidence can be advanced, in that chemisorption of ethane occurs preferentially as ethyl and hydrogen radicals, rather than as methyl radicals, despite the lower bond dissociation energy of the C—C bond. If $Q_{me}$ is the heat of adsorption of methyl, etc., on tungsten, and the preference is entirely due to favourable energetics, we may write

$$2Q_{me} - 83 < Q_{et} + Q_H - 101$$

If, furthermore, $Q_{et}$ is similar to $Q_{me}$ and $Q_H = 72$ kcal mole$^{-1}$

$$Q_{me} < 54 \text{ kcal mole}^{-1}$$

which is satisfied by the observations of Ritchie and Wheeler.

## REFERENCES

1. Brown, S. C., *Basic Data of Plasma Physics*, The Technology Press of the Massachusetts Institute of Technology, and Wiley, New York, 1961.
2. Massey, H. S. W. and Burhop, E. H. S., *Electronic and Ionic Impact Phenomena*, Oxford University Press, 1952.
3. Fomenko, V. S., *Handbook of Thermionic Properties*, Plenum Press Data Division, New York, 1966.
4. Kingdon, K. H., *Phys. Rev.*, **24**, 510 (1924).
5. Richardson, W. O., *Emission of Electricity from Hot Bodies*, Longmans Green, London, 1921.
6. Fowler, R. H., *Statistical Mechanics*, Cambridge University Press, 1936.
7. Hayward, D. O. and Trapnell, B. M. W., *Chemisorption*, Butterworths, London, 1964.
8. Mignolet, J. C. P., *Bull. Soc. Chim. Belg.*, **64**, 126 (1955).

9. Riemann, A. L., *Phil. Mag.*, **20**, 594 (1935).
10. Farragher, A. L., *Ph.D. Thesis*, The University of Aston in Birmingham, 1966.
11. Herron, J. T., Rosenstock, H. M. and Shields, W. R., *Nature*, **206**, 611 (1965).
12. Page, F. M., *Trans. Faraday Soc.*, **56**, 1742 (1960).
13. Eley, D. D., *Disc. Faraday Soc.*, **8**, 34 (1950).
14. Page, F. M., *Nature*, **188**, 1021 (1960).
15. Kay, J. and Page, F. M., *Trans. Faraday Soc.*, **62**, 3081 (1966).
16. Ritchie, B. and Wheeler, R. C., *J. Phys. Chem.*, **70**, 173 (1966).
17. Gaines, A. F., Kay, J. and Page, F. M., *Trans. Faraday Soc.*, **62**, 874 (1966).
18. Kay, J., *Unpublished work*.
19. Goode, G. C., *Unpublished work*.
20. Berry, R. S. and Riemann, C. W., *J. Chem. Phys.*, **38**, 1540 (1963).
21. Cottrell, T. L., *The Strengths of Chemical Bonds*, Butterworths, London, 1958.

# The Correction of the Observed
# Results to Standard Conditions

## 6.1 The Need for Correction

The experimental values for the apparent electron affinity ($E'$) are obtained at various temperatures in the range 1200–2000°K. Since few results lead directly to electron affinities, and most have to be combined with other data, obtained by other techniques at different temperatures, it is necessary to carry out a correction to the standard conditions at some point during the calculations. The mean value of $RT$ in this temperature range is 3 kcal mole$^{-1}$, so that such correction can be very significant, and various ways of making these corrections have been tried. Basically, there are two approaches to the problem, depending upon the theory used to describe the mechanism of ion formation.

## 6.2. The Kinetic Approach

The experimental electron affinity is calculated from the slope of a $\log(j_e p/j_i)$ against $1/T$ plot which corresponds to an apparent electron affinity at a temperature $\bar{T}$, where $\bar{T}$ is the mean value of the temperature measurements. Hence to calculate the electron affinity of the acceptor the apparent electron affinity must be corrected to 0°K.

Page[1] assumed that the ratio $j_e p/j_i$ defined the equilibrium constant $K$ for the reaction

$$A^-(g) + B(g) \rightarrow AB(g) + e$$

Hence $K = A \exp -E'/RT = j_e p/j_i$

$$E_T' = \frac{-R \, d[\log (j_e/j_i)p]}{d(1/T)} = \frac{RT^2 \, d \log K}{dT} = \Delta H_T \qquad (6.1)$$

where $\Delta H_T$ is the enthalpy change for the reaction at temperature $\bar{T}$.

Kirchoff's law gives the relationship between the enthalpy at temperatures $\bar{T}$ and $0°K$ as

$$H_T = H_0 = \int_0^T \Delta C_p \, dT \qquad (6.2)$$

Similarly

$$E_T' = E_0' + \int_0^T \Delta C_p \, dT \qquad (6.3)$$

where $\Delta C_p$ is the difference in heat capacities between reactants and products at constant pressure.

The difference in heat capacities, due to changes in the number of degrees of translational and rotational freedom during reaction, is treated classically as $R/2$ kcal mole$^{-1}$ for each degree of freedom, assuming both reactants and products have attained filament temperature. In most of the corrections made in the early work, the change in *heat capacity at constant volume* ($\Delta C_v$) was used since the temperature dependence of the pre-exponential factor of ion emission was not known, and experience suggested that a factor of $RT$ should be allowed for this (Table 6.1).

It is difficult to determine the extent of the contribution made to $E_T'$ by vibrational terms, since vibrational excitation is normally slower than translational and rotational excitation and only important at elevated temperatures.

In early work, as a first approximation, it was assumed that, at the temperatures used, the stretching modes of vibration were not excited but the bending modes were 'half excited' and, therefore, contributed $\frac{1}{2}RT$ to $E_T'$.

In later work the statistical mechanical expression[2] for the heat capacity at constant pressure due to each vibrational mode was used.

$$\frac{C}{k} = \frac{(\theta/2T)^2}{\sinh^2 (\theta/2T)} \qquad (6.4)$$

where $\theta = h\nu/kT$

$\nu$ is the vibrational frequency

$C$ is the heat capacity due to the vibrational motion at frequency $\nu$.

Table 6.1  Number of Degrees of Freedom *Lost*

| | Rotational | Translational | $\Delta C_v$ | $\Delta C_p$ | |
|---|---|---|---|---|---|
| I  $AB(g) + e(g) \rightarrow AB^-(g)$ | | $+3$ | $+\frac{3}{2}R$ | $\frac{5}{2}R$ | |
| II  (a) $\frac{1}{2}A_2(g) + e(g) \rightarrow A^-(g)$ | | $+3$ | $+\frac{3}{2}R$ | $\frac{5}{2}R$ | |
| (b) $AB(g) + e(g) \rightarrow A^-(g) + B(g)$ (Weak bond) | | $+3$ | | | |
| III  (a) $\frac{1}{2}A_2(g) + e(g) \rightarrow A^-(g)$ | $+1$ | $+\frac{3}{2}$ | $+\frac{5}{4}R$ | $\frac{7}{4}R$ | (A is monatomic) |
| | $\frac{3}{2}$ | $+\frac{3}{2}$ | $0$ | $\frac{1}{2}R$ | (A is polyatomic) |
| (b) $AB(g) + e(g) \rightarrow A^-(g) + B(g)$ (Strong bond) | $+2$ | — | $+R$ | $R$ | (A is monatomic) |
| | — | — | $0$ | $0$ | (A is polyatomic) |
| IV  $AB(g) + e(g) \rightarrow A^-(g) + B_{ads}$ | $+2$ | $+1$ | $+\frac{3}{2}R$ | $2R$ | (A is monatomic) |
| | — | $+1$ | $+\frac{1}{2}R$ | $R$ | (A is polyatomic) |

*Note:* B is taken to be monatomic since experimentally B = Hydrogen or a Halogen atom.

Integrating with respect to $T$ and re-arranging gives

$$\frac{(H_T - H)}{RT} = x \coth x - x \tag{6.5}$$

where $x = 0.720\nu/T$.

Since the vibrational frequencies of negative ions are not, in general, known, it is necessary to assume that the bonds which are not broken during the reaction are the same in the ions as in the parent molecule. If this assumption is made, it is only necessary to identify the vibrational frequencies in the parent molecule which are associated with the broken bond. As vibrational heat capacity corrections are not generally large the approximations used should not cause a serious reduction in accuracy.

Hence if the experimental electron affinity, say for dissociative capture with adsorption, is corrected to $0°K$ by considering changes in the number of degrees of translational, rotational and vibrational freedom during the reaction then

$$E_0' = E_0 - D + Q$$

where $D$ is the bond dissociation energy at $0°K$, since correction of $E_T'$ to $E_0'$ incorporates a correction for $D$.

## 6.3   Transition State Approach

Farragher[3] using the transition state theory derived slightly different equations for correcting the experimental electron affinity to $0°K$. Combining equations (4.15) and (4.26) gives

$$\frac{j_e}{j_i} = \frac{4\pi m_e (kT)^2}{ph^2} \frac{Q_g Q_a^*}{Q_B^* Q_a^*} \exp\left(\frac{-E}{RT}\right) \tag{6.6}$$

As $Q_B^*$ and $Q_a^*$ each possess two translational degrees of freedom they can be replaced by $Q_i$ and $Q_m$; the new *partition functions* representing the vibrational, rotational and electronic contributions only, for the transition state ion and gas phase molecule, respectively.

Noting that $Q_a^*/Q_g = h/(2\pi MkT_g)^{\frac{1}{2}}$, expression (6.6) becomes

$$\frac{j_e}{j_i} = \frac{4\sqrt{2}\,\pi^{\frac{3}{2}}m_e M^{\frac{1}{2}}k^{\frac{5}{2}}T_g^{\frac{1}{2}}T^2}{ph^3} \frac{Q_m}{Q_i} \exp\left(\frac{-E}{RT}\right) \tag{6.7}$$

There are three processes to consider. Direct capture processes, processes where there is dissociation of the parent molecule and processes where there is dissociation with adsorption of a fragment on the filament surface.

## 6.4 Direct Capture Processes

In this case the internal partition functions for the transition state ion and the gas phase molecule are the same except for the electronic contribution. The ground state of the ion will probably be a singlet and the molecule a doublet. Hence $Q_m/Q_i = \frac{1}{2}$ and expression (6.7) becomes

$$\frac{j_e}{j_i} = \frac{2\sqrt{2}\ \pi^{\frac{3}{2}}m_e M^{\frac{1}{2}}k^{\frac{5}{2}}T_g^{\frac{1}{2}}T^2}{ph^3} \exp\left(\frac{-E}{RT}\right) \qquad (6.8)$$

Since $E_T' = R\{d[\log (j_e/j_i)]/d(1/T)\}$ differentiation of (6.8) with respect to $1/T$ gives

$$E_T' = E + 2RT = \Delta H_T \qquad (6.9)$$

where $\Delta H_T$ defines the *enthalpy change* for the reaction

$$A^{-*} = B + e^*$$

In all cases Farragher has neglected the contribution to the apparent electron affinity by vibrational excitation, since this is slow compared with translational and rotational excitation.

## 6.5 Dissociation Without Adsorption

For these processes the only kinetic scheme which is consistent with an energy change of $E - (D + Q)$ is one in which the dissociation is caused by electron capture. Processes in which dissociation precedes electron capture always lead to an $E - (D + Q/2)$ energy balance. If this postulate is correct the only process which is affected by dissociation is the ion forming reaction since adsorption, desorption and electron emission precede capture or are not dependent upon it; hence the transition state ion is of the same form as for direct capture reactions. In this case only the three vibrations associated with the bond which breaks are considered, all others are presumed to be the same in the transition state ion as in the adsorbed molecule. In the transition state, the vibrations associated with the bond which eventually breaks must be almost fully excited, that is, the vibration frequencies are approximately 100 cm$^{-1}$ [4]. Under these circumstances the contribution from the vibrational degrees of freedom to the ratio of the internal partition functions $Q_m/Q_i$ in equation (6.7) cannot be considered as being unity, but must be of the form

$$\frac{Q_m}{Q_i} = \frac{\prod\limits^{3} [1 - \exp(-h\nu_i^{\ddagger}/kT)]}{\prod\limits^{3} [1 - \exp(-h\nu/kT)]} \qquad (6.10)$$

With the transition state frequencies $\simeq 100$ cm$^{-1}$ this approximates to

$$\frac{Q_m}{Q_i} = \left(\frac{h\nu_i^{\ddagger}}{kT}\right)^3 \prod^3 \left[1 - \exp\left(\frac{-h\nu}{kT}\right)\right]^{-1} \qquad (6.11)$$

Hence substituting equation (6.11) into (6.7) and differentiating with respect to $1/T$ gives

$$E_T' = E_0' - RT + N \sum^3 \frac{h\nu}{[\exp(h\nu/kT') - 1]} \qquad (6.12)$$

where $\nu$ = frequency of the vibration in the normal molecule and $T'$ is the vibrational temperature of the adsorbed species. Assuming $T' = T \simeq 1500°$K (the surface temperature) and $\nu \simeq 1000$ cm$^{-1}$, equation (6.12) becomes:

$$E_T' = E_0' - RT + 3 \times 2 \cdot 9 \text{ kcal mole}^{-1} \qquad (6.13)$$

Since $RT \approx 3$ kcal mole$^{-1}$, equation (6.13) approximates to

$$E_T' = E_0' + 2RT \text{ kcal mole}^{-1} \qquad (6.14)$$

### 6.6  Dissociation With Adsorption

In this case the transition state for ion formation must lose two rotational degrees of freedom if it is attached to the surface by the atom X, which is to be adsorbed. The rate determining step is then assumed to be stretching of the A—X bond, where A is the electron acceptor. Hence $Q_m/Q_i$ will now be given by (6.15), assuming as

$$\frac{Q_m}{Q_i} = \frac{\prod^3 [1 - \exp(-h\nu_i^{\ddagger}/kT)]}{\prod^3 [1 - \exp(-h\nu/kT)]} \frac{8\pi^2 kT(I_x I_y)^{\frac{1}{2}}}{\gamma h^2} \qquad (6.15)$$

in the previous case the three vibrations associated with the bond which breaks are excited to $\simeq 100$ cm$^{-1}$. In this case one of these vibrations is rate determining so only 2 vibrations are considered plus the vibration associated with the surface to X bond.

Equation (6.15) approximates to:

$$\frac{Q_m}{Q_i} = \frac{h\nu_1^{\ddagger}}{kT} \frac{1 - \exp(-h\nu_1/kT)}{\prod^3 [1 - \exp(-h\nu/kT)]} \frac{8\pi^2 kT(I_x I_y)^{\frac{1}{2}}}{\gamma h^2} \qquad (6.16)$$

where $\nu_1$ = frequency of surface to X bond vibration

  $\gamma$ = Ehrenfest symmetry factor.

Hence substituting (6.16) into (6.7) and differentiating with respect to $1/T$ gives

$$E_T' = E_0' + N \sum^3 \frac{h\nu}{[\exp(h\nu/kT) - 1]} - \frac{Nh\nu_1}{\exp(h\nu_1/kT) - 1} + RT \qquad (6.17)$$

Assuming $T \simeq 1500°\text{K}$ and $\nu = 1000 \text{ cm}^{-1}$ gives

$$E_T{}' \simeq E_0{}' + 3RT \qquad (6.18)$$

The validity of expressions (6.13) and (6.18) depends upon the vibration frequencies in the molecule, hence expressions (6.12) and (6.17) are more correct. Due to vibrational energy transfer being a slow process the vibrational temperature may lie anywhere between gas temperature and filament temperature. In this case the correction may be anywhere between $-RT$ and $+2RT$ for dissociative capture without adsorption and between $+RT$ and $+3RT$ for dissociative capture with adsorption.

Throughout this book, equations (6.9), (6.14) and (6.18) have been used to correct the experimental electron affinity to $0°\text{K}$.

## REFERENCES

1. Page, F. M., *Trans. Faraday Soc.*, **56**, 1742 (1960).
2. Fowler, R. H. and Guggenheim, E. A., *Statistical Thermodynamics*, Cambridge University Press, 1939.
3. Farragher, A. L., *Ph. D. Thesis*, The University of Aston in Birmingham, 1966.
4. Glasstone, S., Laidler, K. J. and Eyring, H., *The Theory of Rate Processes*, McGraw-Hill, New York and London, 1941.

# Polarization Capture Affinities

## 7.1 The Classification of Negative Ions

In considering and analysing the various results that have been obtained, it is convenient to classify them into four groups according to the type of ion formed.

The first group classes together all ions formed when a free radical accepts an electron to form a singlet ion; the electron is apparently localized as a lone pair and the electron affinity is analogous to a bond energy. This group is discussed in Chapters 10 and 11.

The second group contains doublet ions which are formed from singlet molecules with a delocalized $\pi$ electron system. In this case the energy of the lowest unoccupied orbital in the molecule will be strongly dependent on the internal electrostatic field and the electron affinity may be calculated with reasonable accuracy from simple electrostatic considerations. This group is discussed in Chapters 8 and 9.

A third class of ions are doublet ions which are formed when a singlet molecule, containing no delocalized $\pi$ electron system, captures an electron. The existence of these negative ions cannot be described simply in terms of the orbitals of the uncharged acceptor but must include the distortion of the electron distribution within the ion. These ions are discussed in detail below.

A fourth group contains multiplet ions formed from multiplet acceptors; the only examples found in this work are $S^-$ and $O^-$, and this group is not considered further.

## 7.2 Polarization Affinities

The molecular ion of sulphur hexafluoride $SF_6^-$ has been identified by Fox and Hickam[1], using a mass spectrometer, and by Herron, Rosenstock and Shields[2], who showed that $SF_6^-$ is formed at a hot tungsten surface in the presence of $SF_6$ vapour. In the magnetron there is evidence that at low temperatures (1200–1350°K) processes of ion formation occur in the presence of $SF_6$[3], $CCl_4$[4], $CHCl_3$[4], $CH_2Cl_2$[4], $C_2Cl_6$[4], $BF_3$[5] and $CBr_4$[5] vapours, which are attributed to formation of $SF_6^-$, $CCl_4^-$, $CHCl_3^-$, $CH_2Cl_2^-$, $C_2Cl_6^-$, $BF_3^-$ and $CBr_4^-$ respectively. With $WF_6$[6] and $UF_6$[7] copious ion currents are observed over a temperature range of 1200–1500°K which are similarly attributed to the formation of $WF_6^-$ and $UF_6^-$.

In these ions the negative charge is thought to be distributed between the outer atoms in the molecule; Hush and Segal[8] calculated that for $CFH_3^-$ the negative charge distribution is $0.314e$ and $0.26e$ for the fluorine and hydrogen atoms respectively, with the central carbon atom bearing a slightly positive charge.

Using this calculation as a basis a model may be constructed in which a tetrahedral ion, e.g. $CCl_4$, is represented as a combination of the four states

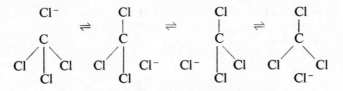

Figure 7.1   The canonical forms of $CCl_4^-$

where the stability of the ion is mainly due to the electrostatic interaction of the negative charge with the polarizable $CCl_3$ group, since the next available orbitals in Cl would require considerable promotion energy.

If the polarizability of the $CCl_3$ group is $\alpha_{CCl_3}$ and the distance between the negative charge and the centroid of the $CCl_3$ group is $r$ then the energy due to electrostatic interaction is given by[9]

$$W = -\frac{1}{2}\frac{e^2}{r^4}\alpha_{CCl_3} \tag{7.1}$$

The effective values of $\alpha_{CCl_3}$ and $r$ may be determined approximately by treating the $CCl_3$ group as three isotropic polarization ellipsoids, each contributing $\frac{1}{4}$ of the molecular (i.e. $CCl_4$) polarizability, with centroids at the mid-point of the C—Cl bond.

Hence
$$W = -\frac{e^2}{2} \sum \frac{\alpha_0}{4r^4} \tag{7.2}$$

$r$ may be calculated in terms of the covalent bond length in the molecule

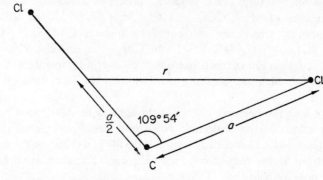

Figure 7.2   Polarization of the C—C— bond by a $Cl^-$ ion

where  $a$  = the covalent bond length C—Cl
      $\alpha_0$ = the molecular polarizability of $CCl_4$

$$W = -\frac{1}{2}\frac{3}{4}\alpha_0 \frac{1}{2\cdot485a^4}e^2 \tag{7.3}$$

$$= -0\cdot1509\frac{e^2\alpha_0}{a^4}$$

Similarly for a planar trigonal ion $XY_3^-$ e.g. $BF_3^-$

$$W = -0\cdot1088\frac{e^2\alpha_0}{a^4} \tag{7.4}$$

for an octahedral ion $XY_6^-$ e.g. $SF_6^-$

$$W = -0\cdot2298\frac{e^2\alpha_0}{a^4} \tag{7.5}$$

and for an ion $X_2Y_6^-$ e.g. $C_2Cl_6^-$

$$W = -0\cdot0724\frac{e^2\alpha_0}{a^4} \tag{7.6}$$

Alternatively, in practical units

$$\text{Trigonal} \quad XY_3 \quad W = -13 \cdot 90 \, \frac{[R]}{a^4}$$

$$\text{Tetrahedral} \quad XY_4 \quad W = -19 \cdot 20 \, \frac{[R]}{a^4}$$

$$\text{Octahedral} \quad XY_6 \quad W = -29 \cdot 30 \, \frac{[R]}{a^4}$$

$$\text{Ethane-like} \quad X_2Y_6 \quad W = -9 \cdot 20 \, \frac{[R]}{a^4}$$

where $[R]$ = molar refraction in ccs

$a$ = X—Y covalent bond length in Å.

and $W$ is in kcal mole$^{-1}$.

Table 7.1 gives the experimental and calculated electron affinities.

In all cases the calculated electron affinity will be a maximum value since the $Y^-$—$XY_3$ bond will be significantly longer than the covalent bond and even the averaged length is likely to be $0 \cdot 15$ Å greater, so that an over-estimation must occur. Nevertheless, the correlation with the experimental electron affinities is quite good for these molecules, with the exception of $BF_3^-$.

In calculating the electron affinities of $CHCl_3$ and $CH_2Cl_2$ it has been assumed, in the absence of charge distribution data, that the negative charge of the anion is equally distributed between the hydrogen and chlorine atoms. This is unlikely to be true, hence the contribution of the

ionic state is over-estimated, which, due to the short C—H bond length, leads to a high value for the calculated electron affinities. If all the bond lengths are taken to be equivalent the calculated electron affinities of $CCl_4$, $CHCl_3$ and $CH_2Cl_2$ are proportional to their molecular polarizabilities, and the agreement between calculated and experimental electron affinities for each compound is similar to that shown in Table 7.1 for $CCl_4$.

Many compounds of the first and second row elements may be treated in this manner, but the extension to the heavy elements such as $WF_6$ and $UF_6$ can only be qualitative because of the unknown contribution of the

Table 7.1

| Compound | Calculated electron affinity (kcal mole$^{-1}$) | Experimental electron affinity (kcal mole$^{-1}$) |
|----------|----------------------------------------------------|-----------------------------------------------------|
| $CCl_4$ | 52·8 | 47·4 |
| $CHCl_3$ | 61·1 | 39·2 |
| $CH_2Cl_2$ | 67·8 | 30·1 |
| $C_2Cl_6$ | 38·9 | 32·8 |
| $BF_3$ | 29·7 | 61·0 |
| $SF_6$ | 41·5 | 33·0 |
| $CBr_4$ | 54·7 | 46·7 |

inner shell electrons of the heavy atom to the total polarizability. There is no data by which to assess this contribution, but the marked increase in the polarizability of the inert gases with increasing atomic weight suggests that it will be substantial. If the bond polarizability in $WF_6$ or $UF_6$ is similar to that in $SF_6$ the contribution to the electron affinity would be of the order of 16 kcal mole$^{-1}$, whereas the contribution due to the polarizability of the central atom, using the data quoted by Syrkin and Dyatkina[10] for the appropriate inert gas, is about 48 kcal mole$^{-1}$, giving a total of 64 kcal mole$^{-1}$ in good agreement with the experimentally determined values of 63·0 and 67·0 kcal mole$^{-1}$ for $WF_6$ and $UF_6$, respectively.

## REFERENCES

1. Fox, R. E. and Hickam, W. M., *J. Chem. Phys.*, **25**, 642 (1956).
2. Herron, J. T., Rosenstock, H. M. and Shields, W. R., *Nature*, **206**, 611 (1965),
3. Kay, J. and Page, F. M., *Trans. Faraday Soc.*, **498**, 1042 (1964).
4. Gaines, A. F., Kay, J. and Page, F. M., *Trans. Faraday Soc.*, **62**, 874 (1966).
5. Kay, J., *Unpublished work.*
6. Goode, G. C., *Unpublished work.*
7. Walters, D., *B.Sc. Thesis*, The University of Aston in Birmingham, 1968.
8. Hush, N. S. and Segal, G. A., *Disc. Faraday Soc.*, April 1968.
9. Castellan, G. W., *Physical Chemistry*, Addison-Wesley, Reading, U.S.A., 1964.
10. Syrkin, J. K. and Dyatkina, M. E., *Structure of Molecules and the Chemical Bond* (transl. abbrev. M. A. Partridge and D. O. Jordon), Butterworths, 1950.

# Direct Capture π Affinities

## 8.1 Experimental Results

The second group of ions which are formed by the direct capture of an electron is much more common than the first group, and all ions of this group contain an extended system of delocalized bonds, together with a number of highly polar bonds. Usually, the neutral precursors are quinones, or nitro and cyano derivatives of benzene, and as such will be recognized as acceptors in charge-transfer complexes, or as the source of the anionic free radicals widely studied by electron spin resonance.

## 8.2 The Additivity of Electron Affinities

The data set out in Table 8.1 shows how the direct capture affinity rises with the number of polar substituent groups in the molecule, and Farragher and Page[1] have analysed this data in terms of group contributions suggesting that a —CN group contributes 12·7 kcal mole$^{-1}$ and the =C(CN)$_2$ group 33·2 kcal mole$^{-1}$ and interpreting those figures in terms of the field of the dipole of the group. This idea of group contributions is not novel, as it was used by Foster[2] in his extensive study of charge-transfer complexes, and the attribution of the effect to the field of the dipole has been exploited by Murrell and Williams[3], but the present data allows a more direct analysis than has hitherto been possible, since the data refers to the gaseous state. Some of the implications of the dipole field model are explored in the next chapter.

4+

Table 8.1

| Compound | Formula | $E_T'$ kcals | $T°K$ | Fila-ment | $E_0$ |
|---|---|---|---|---|---|
| Chloranil | $C_6Cl_4O_2$ | $60.7 \pm 5.9$ | 1350 | Ir | $55.3 \pm 5.9$ |
| Fluorobenzoquinone | $C_6FH_3O_2$ | $54.8$ | 1260 | Ir | $49.8$ |
| Hexafluorobenzene | $C_6F_6$ | $32.8 \pm 1.6$ | 1300 | Ir | $27.6 \pm 1.6$ |
| 1:3:5-trinitrobenzene | $C_6H_3N_3O_6$ | $65.9 \pm 2.4$ | 1370 | Ir | $60.5 \pm 2.4$ |
| $p$-benzoquinone | $C_6H_4O_2$ | $36.7 \pm 0.2$ | 1450 | Ir | $30.9 \pm 0.2$ |
| Tetracyanoethylene | $C_6N_4$ | $72.3 \pm 1.2$ | 1360 | Pt | $66.9 \pm 1.2$ |
| Fluoranil | $C_6F_4O_2$ | $57.7$ | 1370 | Ir | $52.2$ |
| $o$-Phthalonitrile | $C_8H_4N_2$ | $30.4 \pm 2.2$ | 1520 | WC | $24.3 \pm 2.2$ |
| sym tetracyanopyridine | $C_9HN_5$ | $54.0 \pm 1.7$ | 1320 | Ir | $48.7 \pm 1.7$ |
| sym tetracyanobenzene | $C_{10}H_2N_4$ | $54.7 \pm 5.1$ | 1300 | Ir | $49.5 \pm 5.1$ |
| Hexacyanobutadiene | $C_{10}N_6$ | $80.2 \pm 2.4$ | 1370 | Ir | $74.7 \pm 2.4$ |
| 77':88'-tetracyano-quinodimethane | $C_{12}H_4N_4$ | $70.4 \pm 4.4$ | 1330 | Ir | $65.1 \pm 4.4$ |
| Hexacyanobenzene | $C_{12}N_6$ | $62.8 \pm 3.1$ | 1390 | Ir | $57.2 \pm 3.1$ |
| Anthraquinone | $C_{14}H_8O_2$ | $32.1 \pm 2.1$ | 1405 | Ir | $26.5 \pm 2.1$ |
| Fumaronitrile | $C_4H_2N_2$ | $22.1 \pm 2.8$ | 1230 | Ir | $17.2 \pm 2.8$ |

*Notes: Trinitrobenzene.* This result is dubious, since the ion currents were very close to the background current.

*o-Phthalonitrile.* The identification of the process by which this ion is formed as being direct capture is supported by the identical results on filaments of tungsten carbide and iridium.

*Tetracyanoethylene.* Closely similar results were obtained on filaments of iridium, platinum, rhodium, tantalum, molybdenum and tungsten. Rosenstock and Herron have shown by mass spectrometry that the only ion formed is the molecular ion.

*Fumaronitrile.* The result reported here seems too low to be a true direct capture affinity.

A similar additivity has been noted by Naff, Cooper and Compton[4] in a study of the transient negative ion states of aromatic fluorocarbon molecules. Their results, which give essentially the electron affinity of these compounds (for which it is a negative quantity), indicate that this is a linear function of the number of fluorine atoms attached to the ring, and that each fluorine contributes 0·4 eV (9 kcal mole$^{-1}$). Specifically, they find:

Vertical attachment energy

$$= 1·60 - 0·42 \times \text{(number of fluorine substituents)} \quad (8.1)$$

This predicts an electron affinity for hexafluorobenzene of $+0·90$ volt (21 kcal mole$^{-1}$) compared to the value of 27 kcal mole$^{-1}$ in Table 8.1.

## 8.3 The Analysis of the Measured Affinities

The electron affinity $(E)$ may be written generally as if derived from $E_0$ (affinity of basic structure) and $E_g$ (contribution of specific group)

$$E = E_0 + \sum E_g \tag{8.2}$$

Further progress can only yield relative values for $E_g = (E_g{}^1 - E_g{}^2)$ unless some assumption is made about the value of either $E_0$ or $E_g{}^1$. The simplest assumption, made by Farragher and Page, was to set $E_0 = 0$, and to assume that the whole of the measured value of $E$ was due to $E_g$. The error introduced by this assumption is not large, particularly for such highly polar groups as the CN group, but it can lead to errors when the effects of weakly polar groups are being considered.

The results of Compton can, of course, be reconciled with this assumption if the C—H dipole is strongly negative with respect to the C—F dipole, but this is not believed to be the case. A more plausible explanation of the relation found by Compton is that $(E_F - E_H) = 9$ kcal mole$^{-1}$, and $E_0$ has some undetermined value, probably about $-20$ kcal mole$^{-1}$.

It is possible to carry the analysis further, without any arbitrary assignment of a value to $E_0$, if some relation between the $E_g$'s can be found. The discussion given in the next chapter indicates that this can be done on the basis of the bond length and bond moment, and that $E_F = -3E_H$. From this, and the slope of Compton's results, it appears that

$$E_H = -2.2 \text{ kcal mole}^{-1} \qquad E_F = +6.8 \text{ kcal mole}^{-1}$$
$$E_0 = -23.6 \text{ kcal mole}^{-1}$$

The values calculated in the next chapter, however, lead to

$$E_H = -3.6 \text{ kcal mole}^{-1} \qquad E_F = +9.6 \text{ kcal mole}^{-1}$$
$$E_0 = -15 \text{ kcal mole}^{-1}$$

Farragher[5] suggested that some allowance might have to be made for the mutual polarization of adjacent dipoles, and Burdett[6] has explored this possibility using the considerations of Smith[7], and has arrived at a series of 'reduced' dipole moments which she claims give the best fit of all. She found it necessary to include a contribution from $E_0$, or the 'ring affinity' in her nomenclature.

## 8.4 Group Contribution

Whilst it is tempting to try to produce a set of numbers which may be used to correlate all the experimental data, and to predict unknown affinities, the data is still far too scanty to permit this, or to do more than put

Table 8.2

| Group contributions to the electron affinity | |
|---|---|
| Group | Affinity (kcal mole$^{-1}$) |
| C—H | $-2{\cdot}0$ |
| C—F | $4{\cdot}6$ |
| C—Cl | $3{\cdot}8$ |
| C—CN | 17 |
| C—NO$_2$ | 24 |
| =C—(CN)$_2$ | 33 |
| \C=O/ | 20 |

forward some tentative values as in Table 8.2. In advancing such values, we have adhered to the convention of Farragher and Page in assuming that all the measured electron affinity is due to the substituent groups. While this leads to a conflict between our values and those of Compton, it must be remembered that the latter are vertical attachment energies, and that therefore they are upper limits. Other work by Compton suggests that the adiabatic attachment energy for benzene may be 15 kcal lower than the vertical energy, and this difference would account for the supposed 'ring affinity'.

## REFERENCES

1. Farragher, A. L. and Page, F. M., *Trans. Faraday Soc.*, **63**, 2369 (1967).
2. Foster, R., *Tetrahedron*, **10**, 96 (1960).
3. Murrell, J. N. and Williams, D. R., *Proc. Roy. Soc. Series A*, **291**, 224 (1966).
4. Naff, W. T., Cooper, C. D. and Compton, R. N., *Private communication*, 1967.
5. Farragher, A. L., *Ph.D. Thesis*, The University of Aston in Birmingham, 1965.
6. Burdett, M., *Ph.D. Thesis*, The University of Aston in Birmingham, 1968.
7. Smith, J. W., *Electric Dipole Moments*, Butterworths, London, 1955.

# An Electrostatic Approach to the Prediction of π Electron Affinities

## 9.1 Introduction

It has been shown in Chapter 8 that those negative ions which are doublets, with a delocalized $\pi$ electron system, have stabilities which are additive properties of the substituent groups, and the electron affinity $E$ may be written as

$$E = \sum E_i \tag{9.1}$$

where $E_i$ is the group contribution. The experimental data is limited as the study of this type of ion is difficult, owing to the low vapour pressure and thermal instability of most of the parent molecules, but it is precisely this class of compounds which have been widely studied in charge-transfer complexes and for which extended tables of electron affinities have been published. The electron affinities obtained by the magnetron technique are often at variance with those from charge-transfer spectra, and any reconciliation would markedly extend the available data on electron affinities.

The simplest hypothesis is that the group contribution to the electron affinity is due to the dipole of the substituent, which is supported by the correlation between the total dipole moment of the substituent group and the observed electron affinity.

## 9.2 The Field Due to an Assembly of Dipoles

The majority of the molecules showing $\pi$ capture in the magnetron, or acting as acceptors in charge-transfer complexes, are planar, and may be

Table 9.1

| Compound | Electron affinity (kcals) | |
| --- | --- | --- |
| | Magnetron | Charge-transfer[1] |
| Benzoquinone | 30·9 | 14 |
| Anthraquinone | 26·5 | 22 |
| Chloranil | 55·3 | 32 |
| Tetracyanoethylene | 66·5 | 41 |
| Fluorobenzoquinone | 49·8 | 21 |
| 1:3:5-trinitrobenzene | 60·5 | 16 |

ideally represented as a rigid planar assembly of dipoles. The electrostatic energy of a charge and a dipole may readily be calculated for the general case, but for present purposes it will be convenient to assume that the dipoles are placed symmetrically around the normal from the plane of the molecule to the charge.

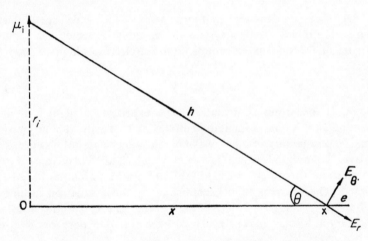

Figure 9.1    The field due to an assembly of dipoles

The electric field at a point x is then given in terms of the radial and azimuthal fields $E_r$ and $E_\theta$ in Figure 9.1.

$$E_r = \frac{2\mu_i \sin \theta}{h^3} \qquad (9.2)$$

$$E_\theta = \frac{\mu_i \cos \theta}{h^3} \qquad (9.3)$$

Resolving these perpendicular and parallel to the line Ox

$$E_\perp = E_\theta \cos \theta - E_r \sin \theta$$

$$= \frac{\mu_i}{h^3} (\cos^2 \theta - 2 \sin^2 \theta)$$

$$E_\parallel = E_\theta \sin \theta + E_\theta \cos \theta$$

$$= 3 \frac{\mu_i}{h^3} \sin \theta \cos \theta \qquad (9.4)$$

The work done in bringing a charge $e$ up to x from infinity is therefore:

$$U = \int_\infty^x E_\parallel \, dx \qquad (9.5)$$

$$= \int_\infty^x \frac{3\mu_i e}{h^3} \sin \theta \cos \theta \, dx \qquad (9.6)$$

$$= 3\mu_i e \int_\infty^x \frac{r_i x}{(r_i^2 + x^2)^{\frac{5}{2}}} \, dx$$

$$= - \frac{\mu_i e r_i}{(r_i^2 + x^2)^{\frac{3}{2}}} \qquad (9.7)$$

If we identify this energy with the group contribution to the electron affinity ($E_i$) when the charge is brought up to 0, in the plane of the molecule, where $x = 0$

$$E_i = \frac{\mu_i e r_i}{r_i^3} = \frac{\mu_i e}{r_i^2} \qquad (9.8)$$

The negative sign has been dropped, because a positive electron affinity represents a decrease in energy.

The apparent lowering of potential energy $E_{ix}$ (a quantity which is relevant to the charge-transfer energy) if the charge is held at a distance $x$ will be given by

$$-U = E_{ix} = \frac{\mu_i e r_i}{(r_i^2 + x^2)^{\frac{3}{2}}} = E_i \left( \frac{r_i^2}{r_i^2 + x^2} \right)^{\frac{3}{2}} = \beta_i E_i \qquad (9.9)$$

### 9.3 The Calculation of Electron Affinities by Electrostatics

The application of equation (9.8) to the calculation of electron affinities has been made by Farragher[2], and Farragher and Page[3], although they used an incorrect coefficient of $\frac{2}{3}$. The problem was also approached, from an entirely different direction by Murrell and Williams[4], who divided the effects of a substituent dipole into a short range, exchange effect, which concerned only the atom to which the substituent group was attached, and a long range inductive effect, which affected all other atoms in the skeleton. Their calculated value for the electron affinity contribution was

$$E_i = 4 \cdot 46\mu$$

compared with (9.8)

$$E_i = 4 \cdot 65\mu \quad \text{for} \quad r_i = 3 \cdot 8 \text{ Å}$$

In applying equation (9.8), the two parameters needed are the group dipole moment, and the distance $r_i$. There has been considerable argument about the values of group moments, and the relative contribution of primary moments and secondary effects to the observed molecular dipole moment. There is still an uncertainty as to the magnitudes of the secondary effects, so for the present calculation the group moments for aromatic compounds listed in Smith[5] will be used, corrected for the C—H dipole = −0·4D. Dipoles will be reckoned positive if their positive pole is directed towards the benzene ring.

Table 9.2    Group Dipole Moments and
Bond Lengths

| Substituent | $\mu$ (Debyes) | $r_{i-x}$ (Å) |
|---|---|---|
| H | −0·4 | 1·08 |
| F | 1·17 | 1·36 |
| Cl | 1·29 | 1·76 |
| Br | 1·31 | 1·94 |
| $NO_2$ | 3·43 | 1·47 |
| CN | 3·63 | 1·40 |
| $CH_3$ | −0·77 | 1·54 |
| $CF_3$ | 2·46 | 1·54 |
| O (ketones) | 2·4 | 1·22 |

There is an inherent difficulty in defining the length $r_i$. Although the molecular geometry is well defined, the point of action of the dipole is not, indeed the implicit assumption of a point dipole though convenient, is inherently false. Smith[5] has quoted the point of action of the aliphatic C—Cl dipole as being either $\frac{2}{3}$ or $\frac{7}{8}$ along the bond, being nearer to the chlorine atom, while Murrell and Williams[4] took the point of action as 1·39 Å from the benzene ring, or 0·79 of the distance along the bond. This fraction is likely to vary with the substituent, and an average value of $\frac{3}{4}$ will be used here. In the case of substituent groups, as distinct from atoms, the point of action will be taken to be the centroid of the group.

The simplest model of a negative ion containing a number of polar groups which can be constructed would place the electron at the centre of the molecule, but this is a physically unrealistic picture. In fact, if the negative ion has two symmetrically placed polar groups, the centre of the molecule is a point of maximum, not minimum, potential. Nevertheless, the calculation of the stability of the negative ion is generally very much easier on this model, and the known structure of charge-transfer complexes indicate that such a symmetrical position is appropriate to the calculation of interaction in these complexes. It is not to be accepted that the charge is located in the centre of the molecule, but rather that the effect of the charge distribution is equivalent to an 'image' electron at that point.

The calculations of Murrell and Williams[4] are based on the dominant contribution of the interaction of a long range force due to the polar group with the atoms of the skeleton other than that to which the group is attached. This model may be adapted to the present considerations by assuming that the added electron in the negative ion has an equal chance of being located on these atoms, and that for a benzene derivative

$$E_i = \frac{1}{5} \sum \frac{\mu_i e^2 \cos \theta}{d_\gamma^2} \equiv \frac{\mu_i e^2}{r_j^2} \qquad (9.10)$$

where $d_\gamma$ is the distance from the point of action of the dipole to the $\gamma$th atom. The parameter $r_j$ now takes the meaning of an effective distance and has been evaluated for various values of the distance from the point of action of the dipole to the benzene ring. The results are shown in Figure 9.2 and the appropriate value of $r_j$ may be read off once the point of action of the dipole has been decided upon.

The results of these calculations are given in Table 9.3, where not only the value of $E_i$, but also that of $E_{ix}$, are given. The value of $r_j$ from Figure 9.2 has been used to calculate $E_i$, but the value of $r_i$ (assuming an image charge on the axis) used to calculate the geometric factor $\beta_i$ of equation (9.9).

It will be seen that the calculated values of $E_i$ are in good agreement with the data available in Table 8.2, except for fluorine and chlorine. It is known that a large part of the dipole moments of fluoro and chlorobenzene are due to a mesomeric moment, and therefore the $\mu_i$ used here may be too large. Murrell and Williams[4] took the effective C—Cl moment to be only 0·75D. It is noteworthy that the $E_i$ (calc) are too large by roughly the same factor (2·09 and 2·37), so that the unobserved experimental value of $E_i$ for Br may be expected to be 3·6. The uncertainty in the dipole moment for the C—H bond suggests that little weight should be attached to the disagreement between observed and calculated values of $E_i$. They agree in that the $E_i$ is small and negative.

4*

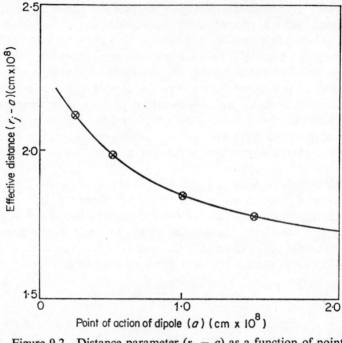

Figure 9.2   Distance parameter $(r_j - a)$ as a function of point
of action of dipole

While the results for substituents in the benzene skeleton are very
satisfactory, the direct application of similar methods of calculation to
the naphthalene and anthracene skeletons are not so satisfactory. The
principal reason for the very large values of $r_j$ which are obtained is that

Table 9.3

| Substitu-ent | Point of action (Å) (0·75 along bond) | $r_i$ (Å) | $r_j$ (Å) | $E_i$ (kcal) Calc. | $E_i$ (kcal) obs. | $\dfrac{E_{ix}}{E_i}$ $(x = 3·2$ Å$)$ |
|---|---|---|---|---|---|---|
| H | 0·82 | 2·22 | 2·74 | −3·6 | −2·0 | 0·185 |
| F | 1·02 | 2·42 | 2·88 | 9·6 | 4·6 | 0·220 |
| Cl | 1·32 | 2·72 | 3·13 | 9·0 | 3·8 | 0·212 |
| Br | 1·45 | 2·85 | 3·24 | 8·5 | — | 0·294 |
| O | 0·92 | 2·35 | 2·81 | 19·3 | 20 | 0·207 |
| $CH_3$ | 1·70 | 3·10 | 3·44 | −4·5 | — | 0·336 |
| $CF_3$ | 1·75 | 3·15 | 3·49 | 13·9 | — | 0·345 |
| CN | 1·97 | 3·37 | 3·67 | 18·1 | 17 | 0·381 |
| $NO_2$ | 2·02 | 3·42 | 3·72 | 16·8 | — | 0·389 |

the contributions of the interaction with the more remote atoms of the skeleton are very small, and these positions are assigned a weight equal to the nearer positions. In order to provide more realistic parameters, the

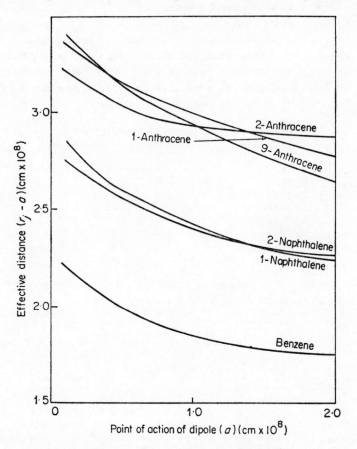

Figure 9.3 The dependence of $(r_j - a)$ on the distance of the dipole in polycyclic aromatic systems

calculated $r_j$ (Figure 9.3) were used as the basis, and were adjusted by the following considerations:

(1) The summations of equation (9.10) can be represented as a core, derived from the previous member of a polycyclic series, together with a small correction term i.e. anthracene ≡ naphthalene + correction.

(2) The most important contributions are those from the two adjacent (ortho) positions in the system.

(3) The effective distance $(r_j - a)$ must fall to zero for large values of $a$.

(4) The $(r_j - a)$ values for the 1 positions in naphthalene and anthracene fell continuously with increasing $a$ and appeared to approach asymptotically the value for benzene.

(5) The $(r_j - a)$ values for the 2 positions were roughly constant and were equal to the value of the corresponding 1 position at $a = 1$ Å. The $(r_j - a)$ values were, therefore, reduced proportionately to agree with the curve for benzene at $a = 3.5$ Å, with the value for the 2 position again being based on that for the position at 1.0 Å. The resulting figures are given in Table 9.4.

Table 9.4    Scaled Parameters $(r_j - a)$ (Å)

| $a$ | 0·5 | 1·0 | 2·0 | 3·0 |
|---|---|---|---|---|
| Benzene | 2·01 | 1·88 | 1·72 | 1·71 |
| 1 Naphthalene | 2·20 | 2·04 | 1·85 | 1·75 |
| 2 Naphthalene | 2·10 | 2·04 | 2·02 | 2·05 |
| 9 Anthracene | 2·35 | 2·16 | 1·87 | 1·75 |
| 1 Anthracene | 2·59 | 2·44 | 2·14 | 2·00 |
| 2 Anthracene | 2·48 | 2·44 | 2·45 | 2·50 |

From this table it is possible to construct Table 9.5 giving the group contributions to the gas phase electron affinity.

Table 9.5

| Position / Group | H | F | Cl | Br | O | CH₃ | CF₃ | CN | NO₂ |
|---|---|---|---|---|---|---|---|---|---|
| Benzene | −3·6 | 9·6 | 9·0 | 8·5 | 19·3 | −4·5 | 13·9 | 18·1 | 16·8 |
| 1 Naphthalene | −3·2 | 8·6 | 8·2 | 7·9 | 18·6 | −4·1 | 12·9 | 17·0 | 15·9 |
| 2 Naphthalene | −3·3 | 8·6 | 7·9 | 7·5 | 18·8 | −3·8 | 12·0 | 15·5 | 14·4 |
| 9 Anthracene | −3·0 | 8·0 | 7·8 | 7·6 | 17·1 | −4·6 | 12·6 | 16·7 | 15·6 |
| 1 Anthracene | −2·5 | 6·7 | 6·5 | 6·3 | 14·5 | −3·3 | 10·5 | 13·9 | 13·0 |
| 2 Anthracene | −2·6 | 6·7 | 6·3 | 5·9 | 14·6 | −3·1 | 9·7 | 12·7 | 11·8 |

## 9.4  Charge-Transfer Spectra

When solutions of certain organic molecules, such as 1:3:5 trinitrobenzene and hexamethyl benzene[6], are mixed, the resulting solution shows in addition to the absorption bands of the components a characteristic absorption band which can be attributed to neither partner, and which is believed to be associated with the transfer of an electron from one molecule, termed the donor, to the other, termed the acceptor. Mulliken described[7,8,9] the donor–acceptor system in terms of a neutral, no bond, ground state (DA) and an excited ionic state ($D^+A^-$). These states interact to produce a stabilized ground state and upper charge-transfer state, with wave functions

$$\psi_0^1 = a_0\psi_0(DA) + b_0\psi_1(D^+A^-)$$
$$\psi_1^1 = a_1\psi_0(DA) + b_1\psi_1(D^+A^-)$$

where $\quad\quad\quad\quad\quad a_0 > b_0 \quad\text{and}\quad a_1 < b_1 \quad\quad\quad\quad\quad$ (9.11)

The absorption bond associated with the transition for $\psi_0^1$ to $\psi_1^1$ will then have an energy given by[10]

$$h\nu_{CT} = I_D(v) - E_A(v) + G_1 + G_0 + X_1 + X_0 \quad\quad (9.12)$$

where $\nu_{CT}$ is the frequency of the band, $I_D(v)$ and $E_A(v)$ are the vertical ionization potential and electron affinity of the donor and acceptor respectively. The $X$ represent the resonance energies of the interaction between the 'no-bond' and 'ionic' states, and the $G$ contain the remaining interaction energy terms. Provided that the inequalities of equation (9.11) are great, the $X$ may be neglected, and equation (9.12) rewritten[11] as

$$h\nu = I - E + C \quad\quad\quad\quad\quad (9.13)$$

where $C$ represents the interaction energy of the excited state relative to the ground state, which is usually taken to be the coulombic (charge–charge) energy.

This equation, and a closely related quadratic form[12] which only differs from (9.13) for ionization potentials less than 7·0 V[13] has been used extensively to correlate and measure ionization potentials, and several attempts have been made to relate the charge-transfer energy to the electron affinities of a series of acceptors[14,15]. The work has been reviewed by Briegleb[1], who later augmented his discussion of electron affinities for which little direct information was available by reviewing a large number of correlations and empirical relations between electron affinities and other parameters, and deriving an extensive table of relative affinities. These differ considerably from the magnetron values, as may be seen from Table 9.1.

It has been observed that while the charge-transfer energy for the complexes between a constant acceptor and a series of donors may be represented by equation (9.13), a better fit is obtained by the empirical relation[15]

$$h\nu = aI - C \quad\quad\quad\quad\quad (9.14)$$

where the constant $a$ has a value rather less than unity. A parallel comparison between $h\nu$ and the values of $E$ obtained by the magnetron method indicates that a similar empirical expression may be used

$$h\nu = bE - C \tag{9.15}$$

with $b$ taking a much lower value around 0·72 (Figure 9.4).

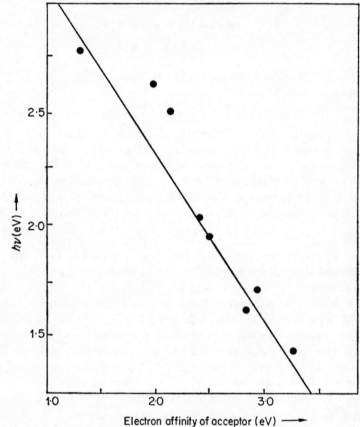

Figure 9.4    Relation between the energy of the maximum of the charge-transfer band and the gas phase electron affinity determined in the magnetron

## 9.5    The Electrostatic Energy of the Complex

If the charge-transfer complex is represented by two planar molecules, each containing a rigid assembly of dipoles, after the model used in a previous section (Figure 9.5), the electrostatic terms which must be considered are:

(1) The charge–charge interaction

(2) The interaction between the charge at A and the dipoles $\mu_a$

(3) The interaction between the charge at B and the dipoles $\mu_a$

(4) The interaction between the dipoles $\mu_a$ and $\mu_b$

(5) The self energy of the overall dipole $A^+B^-$

(6) The reaction field in the solvent due to the dipole $A^+B^-$.

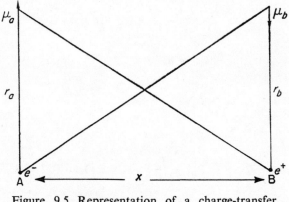

Figure 9.5 Representation of a charge-transfer
complex

If the simplifying assumption that the dipoles $\mu_A$ and $\mu_B$ are the same in the ground, and charge-transfer state is made, the fourth term may be neglected, as it contributes equally in both states. The various terms may now be written down directly to give $C$ in equation (9.13)

$$C = -\frac{e^2}{x} - \sum_a \frac{\mu_a e r_a}{(r_a{}^2 + x^2)^{\frac{3}{2}}} - \sum_b \frac{\mu_b e r_b}{(r_b{}^2 + x^2)^{\frac{3}{2}}} + \frac{\mu^2}{2\alpha} - \frac{f\mu^2}{2(1 - f\alpha)} \quad (9.16)$$

where $\mu$ is the dipole moment of $A^+B^-$, $\alpha$ is the total polarizability of the complex, and $f$ represents the term.

$$f = \frac{1}{a^3} \frac{2\Gamma - 2}{2\Gamma + 1}$$

$a^3$ being the volume of the cavity in the dielectric in which the complex lies and $\Gamma$ the dielectric constant of the solvent. These latter terms have been discussed by Chakrabarti and Basu[16] who have suggested that $\mu$ is usually close to 10D so that the terms represent, respectively, energies of 0·65 and 0·13 eV, assuming the following data as typical; $a = 6\cdot4$ Å, $\Gamma(CHCl_3) = 4\cdot82$, $\alpha = \alpha_A + \alpha_D = 4\cdot8 \times 10^{-23}$ cc. If $x$ is taken to be 3·2 Å, the first term is 4·48 eV, so that $C$ may be written as

$$C = -3\cdot96 - \sum_a \frac{\mu_a e r_a}{(r_a{}^2 + x^2)^{\frac{3}{2}}} - \sum_b \frac{\mu_b e r_b}{(r_b{}^2 + x^2)^{\frac{3}{2}}} \quad \text{(eV)} \quad (9.17)$$

Equation (9.13) now becomes

$$hv = I - E - \sum_a \frac{\mu_a e r_a}{(r_a^2 + x^2)^{\frac{3}{2}}} - \sum_b \frac{\mu_b e r_b}{(r_b^2 + x^2)^{\frac{3}{2}}} - 3 \cdot 96 \qquad \text{(eV)} \quad (9.18)$$

The terms within the summations are, of course, the $E_{ix}$ evaluated in section 9.2 so that

$$hv = I - E + \sum_a E_{ix} - \sum_b I_{ix} - 3 \cdot 96 \qquad (9.19)$$

where $I_{ix}$ denotes the effect of the dipoles of the donor.

If we now write

$$I = I_0 + \sum I_i$$

$$E = \sum E_i$$

$$hv = I_0 + \sum I_i \left(1 - \frac{I_{ix}}{I_i}\right) - \sum E_i \left(1 - \frac{E_{ix}}{E_i}\right) - 3 \cdot 96 \qquad (9.20)$$

Since the value of $E_{ix}/E_i$ does not vary greatly, an average value ($\beta$) may be used so that

$$\sum E_i \left(1 - \frac{E_{ix}}{E_i}\right) = (1 - \beta) \sum E_i = (1 - \beta)E \qquad (9.21)$$

An inspection of Table 9.3 shows that $\beta = 0 \cdot 28$, derived from equation (9.15) represents the mean value well. There will be a similar, but much smaller effect due to the term $\sum I_{ix}$. The term will be much nearer to unity because of the large value of $I_0$ and will be given approximately by

$$a = 1 - \frac{\beta \sum I_i}{I_0 + \sum I_i} \qquad (9.22)$$

so that $a \approx 0 \cdot 94$ for the series of methylbenzenes, since $\sum I_i \approx 0 \cdot 25 I_0$.

The agreement of this value with the empirically based coefficient in equation (9.14) is interesting, but not very important because of the closeness to unity, and because the empirical value is not well established. The empirical coefficient of the electron affinity is, however, significantly different from unity and, despite the paucity of experimental data, bears a more detailed examination.

If the expression for the charge-transfer energy is written in the form

$$hv = aI - \sum \beta_i E_i - C \qquad (9.23)$$

Table 9.6

| Substituent Position | H | O | F | F | Cl | Cl | Br | Br | CH$_3$ | CF$_3$ | CN | NO$_2$ |
|---|---|---|---|---|---|---|---|---|---|---|---|---|
| Benzene | −0.13 | 0.66 | 0.32 | 0.14 | 0.29 | 0.13 | 0.26 | 0.12 | −0.13 | 0.40 | 0.49 | 0.45 |
| 1 Naphthalene | −0.10 | 0.60 | 0.27 | 0.12 | 0.25 | 0.11 | 0.23 | 0.10 | −0.11 | 0.35 | 0.44 | 0.40 |
| 2 Naphthalene | −0.11 | 0.60 | 0.27 | 0.12 | 0.25 | 0.11 | 0.23 | 0.10 | −0.11 | 0.35 | 0.44 | 0.40 |
| 1 Anthracene | −0.08 | 0.58 | 0.21 | 0.10 | 0.19 | 0.09 | 0.18 | 0.08 | −0.09 | 0.28 | 0.35 | 0.32 |
| 2 Anthracene | −0.07 | 0.36 | 0.16 | 0.08 | 0.14 | 0.06 | 0.13 | 0.06 | −0.07 | 0.21 | 0.26 | 0.24 |
| 9 Anthracene | −0.17 | 0.59 | 0.27 | 0.12 | 0.25 | 0.11 | 0.23 | 0.11 | −0.12 | 0.36 | 0.45 | 0.41 |

then the terms involving the ionization potential and charge–charge interactions may be combined, if a constant donor is considered, so that

$$h\nu = A - \sum \beta_i E_i \qquad (9.24)$$

the terms $\beta_i E_i$ are given in Table 9.6 for the substituent groups of Table 9.3, considering in turn the benzene skeleton, the 1 and 2 position of naphthalene and the 1, 2 and 9 positions of anthracene. The values are given in electron volts to facilitate comparison with the charge-transfer energies. Two values are given for the halogen substituents, corresponding to the $E_i$ calculated from the dipole moment, and to a value 2·2 times less than this, corresponding to the group contribution to the gas phase electron affinity.

These contributions may be summed to calculate the charge-transfer energy of the complexes of a single donor with a variety of acceptors. This has been done in Table 9.7 for the complexes of pyrene, one of the more widely used donors. The ionization potential has been taken as 7·53 eV so that

$$h\nu = 3·62 - \sum \beta_i E_i \qquad (eV) \qquad (9.25)$$

The agreement, with the exceptions of nitrobenzoquinone, tetra-cyanobenzoquinone, 1:4 naphthoquinone, pentacyanotoluene and tetra-cyanoxylene is everywhere reasonable, and in many cases very good. The greatest deviations occur for small values of $h\nu$ where the quadratic expression is to be preferred and the values of the charge-transfer energy have also been calculated from the empirical relation

$$h\nu = \left(3·39 - \sum \beta_i E_i\right) + \frac{0·5}{(3·39 - \sum \beta_i E_i)} \qquad (9.26)$$

which fits slightly better below $h\nu = 2·00$ eV. The objective of Table 9.7 is not, however, to produce the best empirical fit to the experimental observation, by the use of adjustable parameters, but rather to assess an absolute calculation of the charge-transfer energy based on an electrostatic model.

The very close correspondence between $h\nu_{obs}$ and $h\nu_{calc}$ is most satisfactory, particularly so when it is remembered that all the data used in the calculation were obtained by completely independent methods, with the sole exception of the effective dipole moments of the halogens which were taken to be 0·45 of the accepted value, in order to bring the gaseous electron affinities calculated electrostatically in line with the observed values. Even here the data owes nothing to charge-transfer spectra, but does reflect on the model used.

Table 9.7

| Compound | Observed | $h\nu$ (eV) Calc. (linear) | Calc. (quadratic) |
|---|---|---|---|
| *p*-benzoquinone | 2·77 | 2·79 | 2·76 |
| Nitrobenzoquinone | 1·88 | 2·21 | 2·23 |
| Cyanobenzoquinone | 2·24 | 2·17 | 2·20 |
| Fluorobenzoquinone | 2·63 | 2·51 | 2·50 |
| Chlorobenzoquinone | 2·53 | 2·53 | 2·52 |
| Bromobenzoquinone | 2·50 | 2·55 | 2·54 |
| Methylbenzoquinone | 2·84 | 2·79 | 2·76 |
| Dinitrobenzoquinone | 1·65 | 1·63 | 1·76 |
| Dichlorobenzoquinone | 2·33 | 2·27 | 2·29 |
| Dicyanobenzoquinone | 1·73 | 1·55 | 1·70 |
| Dibromobenzoquinone | 2·20 | 2·29 | 2·30 |
| Trichlorobenzoquinone | 2·10 | 2·01 | 2·06 |
| Tetracyanobenzoquinone | 1·10 | 0·31 | — |
| Tetrachlorobenzoquinone | 2·03 | 1·75 | 1·85 |
| Tetrabromobenzoquinone | 2·03 | 1·80 | 1·89 |
| Trichloromethylbenzoquinone | 2·29 | 2·01 | 2·06 |
| Dibromodimethylbenzoquinone | 2·46 | 2·29 | 2·30 |
| Dichlorodicyanobenzoquinone | 1·45 | 1·30 | 1·54 |
| | | | |
| 1:4 Naphthoquinone | 2·60 | 3·05 | 3·00 |
| Dichloronaphthoquinone | 2·53 | 2·62 | 2·60 |
| Dibromonaphthoquinone | 2·40 | 2·63 | 2·61 |
| Dimethylnaphthoquinone | 3·00 | 3·06 | 3·01 |
| Chlorobromonaphthoquinone | 2·51 | 2·63 | 2·61 |
| | | | |
| 9:10 Anthraquinone | 2·92 | 3·03 | 2·98 |
| Chloroanthraquinone | 2·89 | 2·86 | 2·82 |
| Methylanthraquinone | 2·99 | 3·04 | 2·98 |
| Nitroanthroquinone | 2·72 | 2·63 | 2·61 |
| Bromoanthraquinone | 2·87 | 2·87 | 2·83 |
| Dichloroanthroquinone | 2·85 | 2·69 | 2·66 |
| Dibromoanthroquinone | 2·79 | 2·71 | 2·68 |
| Trichloroanthroquinone | 2·82 | 2·52 | 2·51 |
| | | | |
| Trinitrobenzene | 2·79 | 2·66 | 2·00 |
| Pentacyanotoluene | 2·32 | 1·40 | 1·60 |
| Tetracyanoxylene | 2·83 | 1·96 | 2·02 |
| Hexacyanobenzene | 1·95 | 0·66 | 1·59 |
| Tetracyanobenzene | 2·51 | 1·70 | 1·80 |

Since there is such very good agreement between the absolute calcula-tion of the charge-transfer energy and the observed value, it is reasonable to accept the model proposed as being basically correct, and to enquire about the correlation of charge-transfer data with other estimates of the electron affinity. It is immediately apparent from Table 9.7 and Figure 9.4 that the charge-transfer data are quantitatively expressed by

$$hv = A - \sum \beta_i E_i$$

but are only poorly reproduced by

$$hv = A - b \sum E_i$$

or
$$hv = A - bE$$

The accuracy of the latter equation depends on the divergence of the $\beta_i$ from a uniform value $b$. If the divergence is unimportant then either equation may be used, and a satisfactory correlation between phenomena related to the electron affinity will be found, so that all the correlations used by Briegleb[1] will still be valid if it is remembered that his values of '$EA$' are $0.70E$ ($bE$). On the other hand, there are many instances where it is improper to use an averaged value of $\beta_i$ so that the correlations may fail. In particular, it will not be possible to obtain electron affinities from charge-transfer data, although it may be possible to determine the $\beta_i E_i$ contribution of a group and thence estimate the gas phase affinity. Also, it will not be valid to compare charge-transfer energies if the complexes have differing structures, e.g. to base a scale of electron affinities for substituted benzoquinones on the charge-transfer energy for the complexes of the halogen molecules[13]. Estimates of electron affinities derived from charge-transfer spectra or similar correlations must therefore be viewed with caution and can only be considered to have qualitative significance.

## REFERENCES

1. Briegleb, G., *Angew Chem.* (Intern. ed.), **3**, 617 (1964).
2. Farragher, A. L., *Ph.D. Thesis*, The University of Aston in Birmingham, 1966.
3. Farragher, A. L. and Page, F. M., *Trans. Faraday Soc.*, **63**, 2369 (1967).
4. Murrell, J. N. and Williams, D. R., *Proc. Roy. Soc.*, *Series A*, **291**, 224 (1966).
5. Smith, J. W., *Electric Dipole Moments*, Butterworths, London, 1955.
6. Briegleb, G. and Czekalla, J., *Z. Phys. Chem.* (Frankfürt), **24**, 37 (1960).
7. Mulliken, R. S., *J. Amer. Chem. Soc.*, **72**, 600 (1950).
8. Mulliken, R. S., *J. Amer. Chem. Soc.*, **74**, 811 (1952).
9. Mulliken, R. S., *J. Phys. Chem.*, **56**, 801 (1952).
10. Mulliken, R. S. and Person, W. B., *Ann. Rev. Phys. Chem.*, **13**, 107 (1962).
11. Ham, J. S., McConnell, H. and Platt, J. R., *J. Chem. Phys.*, **21**, 66 (1953).
12. Franklin, J. L., Hastings, S. H., Matsen, F. A. and Schiller, J. C., *J. Amer. Chem. Soc.*, **75**, 2900 (1953).
13. Briegleb, G. and Czekalla, J., *Z. Electrochem.*, **63**, 6 (1959).
14. Batley, M. and Lyons, L. E., *Nature*, **196**, 573 (1962).
15. Briegleb, G. and Czekalla, J., *Angew Chem.*, **72**, 401 (1960).
16. Basu, S. and Chakrabarti, S. K., *Trans. Faraday Soc.*, **60**, 465 (1964).

# σ-Affinities

## 10.1 Introduction

The capture of an electron by a doublet free radical, leading to the formation of a singlet negative ion, may be called a σ-process, since the electron held in a lone pair may be likened to an atom held in a σ-bond. It might be supposed that the additional electron could be delocalized, but, as will be seen below, there is a considerable body of evidence that such an electron remains as a lone pair associated with the atom at which the original free valence lay. The electron may be regarded as an ordinary chemical reagent substituting for another atom.

$$AB + e \rightarrow A^- + B$$

The σ-electron affinity of A is virtually independent of the substitution of groups in A, just as the A—B bond energy is almost independent. It does depend strongly on the nature of the centre at which substitution occurs, and on the hybridization of that centre in a manner completely analogous to the dependence of the bond energy; in fact there is such a close parallel that the σ-affinities may be regarded as the A—$e$ bond energies.

## 10.2 Electron Affinities of Atoms and Radicals

If the σ-affinity is to be regarded as a bond energy, it is proper to enquire if there is any connexion between the electron affinity of a radical XY and that of the atomic centre of acceptance Y.

Starting with the radical, the atomic negative ion $Y^-$ may be formed by two processes:

(1)  Form the radical ion $XY^-$     $\Delta H = -E_r$
     Dissociate the radical ion     $\Delta H = D_i$
(2)  Dissociate the radical         $\Delta H = D_r$
     Form the atomic ion            $\Delta H = -E_a$

$$X + Y + e \xrightarrow{\ -E_a\ } X + Y^-$$

$$\uparrow D_r \qquad\qquad D_i \uparrow$$

$$XY + e \xrightarrow{\ -E_r\ } XY^-$$

Hence                $D_r - E_a = D_i - E_r$

or                         $E_r = E_a + (D_i - D_r)$                    (10.1)

As the bond dissociation energies in the radical and the ion will be nearly the same, unless there is a great disparity in the electronegativities of X and Y, or a change in the hybridization of Y on dissociation, the term within brackets will tend to zero. If the restriction that Y does not change in hybridization is imposed, that is, that Y remains in the same valence state throughout the cycle, then only a great difference in electronegativity will cause the term $(D_i - D_r)$ in (10.1) to become significant, and experience suggests that only when X is fluorine is such a difference important. For all other substituents, the term is small, and we may identify the $\sigma$-electron affinity of a radical with the electron affinity of the acceptor atom *in the same valence state*.

## 10.3  Valence States and Promotion Energies

An atom, when free, will exist in a well-defined state, whose energy level is equally well defined. These energy levels are known with precision from spectroscopic observations, and correspond to states in which the electrons occupy orbitals which have spherical, threefold or fivefold symmetry. When these atoms enter into chemical bonding, the bonds formed have, in general, different symmetries around the atom, and the bonding electrons are described as occupying orbitals which are formed by combinations, or hybrids, of the original (free atom) orbitals, which satisfy the new requirements of molecular symmetry. For example, the ground state of the carbon atom is a $^3P$ state, with the electronic configuration $s^2pp$. This is a divalent state, with the two bonds at right angles, and the lowest state with four unpaired electrons is the *sppp* state, which requires 195 kcal more energy than the divalent state. The bonds formed from the *sppp* state are not equivalent, or arranged tetrahedrally, but a suitable set of four equivalent tetrahedral orbitals can be formed from the *s* and *p* orbitals, which results in an increase in the strength of the bonds formed, and an overall decrease in the potential energy of the system, so that a greater stability ensues. This set of hybrid orbitals are denoted $sp^3$ (from their components) or *te* (from their symmetry) orbitals, and the carbon atom is said to be in an $sp^3$ or *tetetete* valence state. The energy of a valence state, for an atom may have many, does not correspond to that of any observable spectroscopic state unless the spectroscopic state is also a valence state as with H or Cl.

The realization that a change in the hybridization of the valence state during the breaking of a chemical bond would lead to a change in the energy of the system, manifest as a variation in the bond energy, led to several attempts to estimate the energies of the valence states. These attempts are summarized in Cottrell's monograph[1].

## 10.4 The Method of Hinze and Jaffe

Hinze[2], and later Hinze and Jaffe[3,4], set out to evaluate the electronegativities of elements in their valence states, and as part of this programme calculated the promotion energies of a large number of atoms from their spectroscopic ground states to their valence states. These calculations followed the method of Mulliken[5], wherein a suitable set of atomic orbitals are chosen which will combine to give a set of equivalent orbitals of the requisite symmetry. This choice may be made by a variational procedure, but can be readily achieved by the application of group theory and molecular symmetry. The conditions for equivalence, normalization and orthogonality will then lead to a definition of the hybrid orbitals in terms of the generating atomic orbitals, from which the energy of the valence state and the promotion energy may be obtained. These two energies are equivalent provided that, in the determination of the Slater–Condon parameters needed for the evaluation of the valence state energy, the zero of the energy scale was chosen as the energy of the spectroscopic ground state.

This procedure is practicable for neutral atoms and positive ions, but as no spectra have been observed for negative ions, their promotion energies may not be obtained in the same way, since the Slater–Condon parameters are not available. Rohrlich[6], however, demonstrated that there was a straight-line relation between the corresponding spectroscopic states of successive isoelectronic ions, and Hinze used a similar extrapolation to derive the promotion energies for the negative ions.

The promotion energies obtained by these procedures are set out in Table 10.1 which has been abridged from the thesis of Hinze.

Table 10.1  The Promotion Energies of Atoms and Ions in kcal mole$^{-1}$

|  | Li |  | Na |  | K |  | Rb |  |
|---|---|---|---|---|---|---|---|---|
| *s* |  | 0 |  | 0 |  | 0 |  | 0 |
| *p* |  | 42·6 |  | 48·5 |  | 37·2 |  | 36·4 |
|  | Li$^-$ | Be | Na$^-$ | Mg | K$^-$ | Ca | Rb$^-$ | Sr |
| *sp* | 25·0 | 77·6 | 22·7 | 71·8 | 56·0 | 49·6 | 10·3 | 47·0 |
| *pp* | 52·7 | 165·2 | 42·0 | 148·1 | 35·0 | 112·1 | 28·2 | 102·9 |
| *didi* | 18·7 | 62·8 | 18·9 | 63·6 | 40·0 | 44·1 | 12·6 | 42·2 |
| *di*$\pi$ | 38·8 | 121·4 | 32·4 | 110·0 | 45·5 | 80·9 | 19·3 | 74·9 |
| *trtr* | 31·4 | 100·3 | 27·5 | 93·6 | 41·2 | 68·0 | 17·3 | 63·5 |
| *tr*$\pi$ | 43·5 | 136·0 | 35·6 | 122·7 | 42·0 | 91·3 | 22·3 | 84·3 |
| *tete* | 37·3 | 117·7 | 31·4 | 107·9 | 41·5 | 79·5 | 19·8 | 73·7 |
|  | Be$^-$ | B | Mg$^-$ | Al | Zn$^-$ | Ga | Cd$^-$ | In |
| *spp* | 66·6 | 129·6 | 59·8 | 112·0 | 84·0 | 134·6 | 93·5 | 131·3 |
| *ppp* | 139·4 | 279·8 | 137·1 | 245·1 | 184·5 | 288·2 | 186·3 | 269·3 |
| *didi*$\pi$ | 54·6 | 10·93 | 52·5 | 99·6 | 75·6 | 117·4 | 87·4 | 121·2 |
| *di*$\pi\pi$ | 103·0 | 204·7 | 98·5 | 178·5 | 134·3 | 211·4 | 139·9 | 200·3 |
| *trtrtr* | 50·5 | 102·5 | 50·1 | 95·5 | 72·8 | 118·4 | 85·4 | 117·8 |
| *trtr*$\pi$ | 85·5 | 170·6 | 82·4 | 150·9 | 113·8 | 180·4 | 121·7 | 172·8 |
| *tetete* | 75·7 | 151·9 | 73·7 | 135·9 | 102·8 | 165·1 | 112·1 | 158·2 |
|  | B$^-$ | C | Al$^-$ | Si | Ga$^-$ | Ge | In$^-$ | Sn |
| *sppp* | 117·4 | 195·6 | 89·1 | 143·5 | 94·6 | 151·6 | 105·2 | 148·0 |
| *didi*$\pi\pi$ | 93·3 | 165·9 | 74·8 | 124·9 | 97·7 | 146·8 | 97·5 | 139·2 |
| *trtrtr*$\pi$ | 85·6 | 156·0 | 70·1 | 118·6 | 98·7 | 145·2 | 95·0 | 136·2 |
| *tetetete* | 81·6 | 151·0 | 67·7 | 115·6 | 99·2 | 144·4 | 93·7 | 134·7 |
|  | C$^-$ | N | Si$^-$ | P | Ge$^-$ | As | Sn$^-$ | Sb |
| *s$^2$ppp* | 15·7 | 25·0 | 17·2 | 19·2 | 21·0 | 20·5 | −7·0 | 6·7 |
| *sp$^2$pp* | 213·4 | 329·6 | 112·2 | 182·0 | 81·1 | 172·6 | 48·1 | 152·8 |
| *di$^2$di*$\pi\pi$ | 114·6 | 177·3 | 64·7 | 100·6 | 51·2 | 96·6 | 20·6 | 79·7 |
| *didi*$\pi^2\pi$ | 189·3 | 296·7 | 107·8 | 171·2 | 79·0 | 164·6 | 47·0 | 145·2 |
| *tr$^2$trtr*$\pi$ | 136·8 | 213·4 | 78·6 | 123·2 | 60·3 | 118·3 | 29·2 | 100·7 |
| *trtrtr*$\pi^2$ | 181·2 | 285·8 | 106·3 | 168·4 | 78·3 | 161·9 | 46·5 | 142·6 |
| *te$^2$tetete* | 145·9 | 228·8 | 85·1 | 133·6 | 64·6 | 128·6 | 33·5 | 110·5 |
|  | N$^-$ | O | P$^-$ | S | As$^-$ | Se | Sb$^-$ | Te |
| *s$^2$p$^2$pp* | 8·1 | 12·4 | 4·9 | 7·1 | 4·8 | 8·9 | 4·8 | 10·1 |
| *sp$^2$p$^2$p* | 272·0 | 391·3 | 154·2 | 218·2 | 109·5 | 189·5 | 94·6 | 160·4 |
| *di$^2$di$^2$*$\pi\pi$ | 8·1 | 12·4 | 4·9 | 7·1 | 4·8 | 8·9 | 48·0 | 101·2 |
| *di$^2$di*$\pi^2\pi$ | 140·1 | 201·8 | 79·5 | 122·7 | 57·2 | 99·2 | 49·7 | 85·2 |
| *didi*$\pi^2\pi^2$ | 248·4 | 358·8 | 143·0 | 201·7 | 104·6 | 178·7 | 89·4 | 151·1 |
| *tr$^2$tr$^2$tr*$\pi$ | 96·1 | 138·7 | 54·7 | 77·5 | 39·7 | 69·1 | 34·7 | 60·2 |
| *tr$^2$trtr*$\pi^2$ | 173·5 | 250·5 | 99·5 | 140·5 | 72·5 | 124·5 | 62·4 | 106·1 |
| *te$^2$te$^2$tete* | 134·2 | 193·7 | 76·7 | 108·5 | 55·9 | 96·5 | 48·4 | 82·9 |
|  | O$^-$ | F | S$^-$ | Cl | Se$^-$ | Br | Te$^-$ | I |
| *s$^2$p$^2$p$^2$p* | 0·25 | 0·4 | −0·1 | 0·8 | 1·5 | 3·5 | 3·8 | 7·2 |
| *sp$^2$p$^2$p$^2$* | 346·7 | 481·8 | 185·1 | 248·1 | 193·2 | 252·5 | 145·7 | 234·5 |

106

In constructing this table, only states of the common multiplicity have been considered, e.g. states of carbon with two singly occupied hybrid orbitals only (carbenes) have not been included, nor have those with all electrons in pairs. The original tables do not include the promotion energies for atoms and ions with open $d$ shells. The calculations for these are more complex, and the results cannot be presented as succinctly as those with closed $d$ shells set out above.

## 10.5   The Prediction of Electron Affinities

If equation (10.1) is a good approximation, and the bond energy term vanishes, the electron affinity of a radical will be given by the electron affinity of the atom in the appropriate valence state. Table 10.1 shows the promotion energies of the atom and ion from their ground states, and if these are denoted by $P_a$ and $P_i$ respectively the electron affinity of the atom in the valence state $(E_v)$ will be greater than that in the ground state $(E_g)$ by the amount $(P_a - P_i)$

$$E_v = E_g + (P_a - P_i) \tag{10.2}$$

There are, of course, a number of possible atomic valence states, as there are also for the ion, but some transitions between the atomic and ionic states are less likely than others, the most probable being those in which no rehybridization occurs, the extra electron entering a vacant orbital to form a lone pair. Such transitions are illustrated in Figures 10.1 and 10.2 where correlation diagrams for carbon and nitrogen are shown. It is apparent that carbon, in the *trtrtrπ* state, with three trigonal and one π orbital, can accept an electron into either the trigonal or the π orbital, to give the *tr²trtrπ* or the *trtrtrπ²* states of the ion. If rehybridization can occur, the *te²tetete* state of the ion is possible, and in the extreme, ionic states with two lone pairs, such as *tr²tr²tr*, might merit consideration.

The factors which limit the choice of valence states of atom and ion are threefold: the chemical structure of the radical, the presumed geometry and the energy. In the example cited, the trigonal state can represent a phenyl or an ethylenic carbon atom, in which one trigonal orbital is free, and the other orbitals are utilized in the molecular bonding. The only possible ion is therefore *tr²trtrπ*, with the extra electron going into the free trigonal orbital. On the other hand, the same atomic valence state could represent a planar methyl radical, using the three trigonal orbitals for bonding, and having a free π orbital, in which case the ion formed would be in the *trtrtrπ²* state. The chemical structure can therefore be of material importance in deciding which valence states merit consideration.

The chemical structure of the whole radical can also be of relevance. The methyl and benzyl radicals have, so far as the active centres are concerned, the same structure, but the repulsion of the trigonal orbitals by the single $\pi$ electron will cause the methyl radical to take a slightly

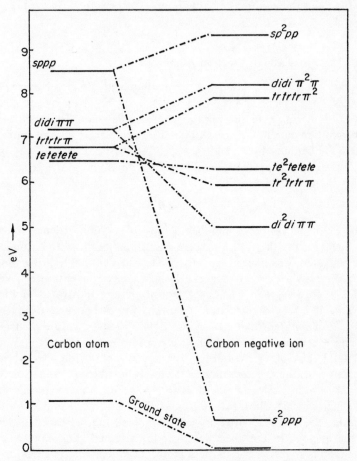

Figure 10.1 Correlation diagram to determine the electron affinities of a carbon atom in various valence states

pyramidal shape. The resonance stabilization of the benzyl radical will lock it into the planar form. The addition of a second electron in the $\pi$ orbital will materially increase the repulsion, and the methyl ion, at least, will be of pyramidal shape, so that the valence state $trtrtr\pi^2$, based as it is on a planar geometry, will be a poor approximation, yet the distortion will not be so great as to merit description by the tetrahedral *te²tetete* state. Some allowance can be made for this distortion, either by expressing the state as a hybrid of hybrids or by allowing for the energy required to bend the skeleton, which may be estimated from the force constants of the appropriate infra-red vibration bands. The effects are

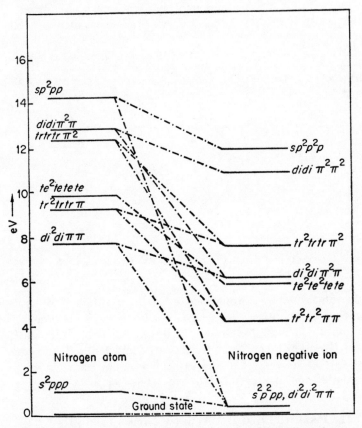

Figure 10.2 Correlation diagram to determine the electron affinities of a nitrogen atom in various valence states

clearly seen in Figure 10.1, where the electron affinity is 1 kcal if the ion is in the planar *trtrtrπ²* state, but 36 kcal if it is in the rehybridized *te²tetete* state. The observed value (Table 11.2) of 24 kcal clearly indicates a considerable degree of distortion towards the tetrahedral state, and the lower value for benzyl (19 kcal) in the same table could reflect the lesser degree of distortion.

## 10.6  Atomic Electron Affinities

Having selected the appropriate valence states for atom and ion, and the relevant promotion energies, the only additional information required to predict the electron affinity from equation (10.2) is the electron affinity of the atom in the ground state. These atomic electron affinities are known experimentally for a few atoms only, but reasonably reliable interpolations based on careful theoretical discussions are available. Table 10.2

Table 10.2  Atomic Ground State Electron Affinities (kcal mole$^{-1}$)

| Element | Affinity | Element | Affinity | Element | Affinity |
|---------|----------|---------|----------|---------|----------|
| H  | 17·3  | P   | 17·8 | Zn | −20·8 |
| Li | 18·9  | S   | 47·7 | Ga | 4·2   |
| Be | −4·4  | Cl  | 85·1 | Ge | 27·7  |
| B  | 7·6   | K   | 0·0  | As | 15·0  |
| C  | 25·8  | Ca  | 2·5  | Se | 50·7  |
| N  | 1·2   | Sc  | 10·6 | Br | 81·9  |
| O  | 33·9  | Ti  | 28·8 | Rb | 0·0   |
| F  | 80·2  | V   | 27·4 | Sr | 2·3   |
| Ne | −13·1 | Cr  | 81·6 | In | 4·6   |
| Na | 10·8  | Mn  | 27·0 | Sn | 23·1  |
| Mg | −7·4  | Fe  | 72·2 | Sb | 25·4  |
| Al | 12·0  | Co  | 75·9 | Te | 53·0  |
| Si | 33·7  | Ni, Cu | — | I  | 74·0  |

lists the values used by Hinze, which follow those of Edlen[7] for the lighter elements, and of Ginsberg[8] for the heavier. For carbon the value is that of Branscomb and Smith[9], for Group VIa, Branscomb et al.[9,10], or Doerffel[11], and for Group VIIIa those of Cubicciotti[12] and Bailey[13]; the latter are slightly higher than the latest experimental values from the schools of Branscomb[14] and of Berry[15].

## 10.7  The Predicted Electron Affinities of Groups I and II

In order to make use of the method of Hinze and Jaffe to correlate the observed electron affinities, it has been applied to the hydrides of the elements in the first two rows of the periodic table by Gaines and Page[16], and is extended here to cover all elements for which valence state promotion energies are available. The various groups, for which the ions will involve similar valence states, will be discussed in order.

No radicals can be formed from the elements of the first group (Ia) unless valence states of an improbable degree of excitation are involved.

The only ions in this group will therefore be the atomic ions, whose stabilities are given in the previous section. The elements of group Ib, which have incomplete $d$ shells, are not readily susceptible to the treatment of Hinze and Jaffe, and no promotion energies are available.

In Group IIa, the valence states of beryllium and magnesium in their hydrides will be *didi*, and in the ion *di²di*, though there is also the possibility of a triplet ion (*trtrtr*). The values quoted for calcium and strontium are estimates made by Hinze, as the promotion energies for these ions cannot be calculated. No experimental data are available with which to compare these predicted affinities.

Table 10.3   The Electron Affinities of Group IIa

| Transition | | BeH | MgH | CaH | SrH |
|---|---|---|---|---|---|
| Atom | Ion | | | | |
| *didi* | *di²di* | 23 | 25 | 24 | 21 |
| *didi* | *trtrtr* | 19 | 21 | | |

## 10.8   The Predicted Electron Affinities of Group IIIa

The increased number of electrons in this group makes possible the choice of several different valence states. The lowest lying of these states is the unhybridized $s^2p$ state of the atom yielding a $s^2pp$ state of the ion, which would be just stable, but the observed ionization potential of BH (8·7 or 10·1 eV) suggests that the valence state should be *di²di* for the radical, which would be expected to have a lone pair ionization potential of 9·6 eV, rather than $s^2p$, for which an ionization potential of 14·0 eV would be expected. Similar arguments apply to the other monohydrides. Strictly speaking the negative ions formed from these hydrides will be in a doublet state ($s^2pp$ or *di²diπ*) so that the estimation is not valid, but the affinities are included for comparison.

The doubly bonded states of this group may be derived either from the trigonal (*trtrtr*) state, or the digonal (*didiπ*) state, and since the promotion energies of these two states are close, a number of stable negative ions may be formed. If the structure of the radical is such that only the digonal state is possible, for example in BO (or $B_2H$), the resultant ion can only be in a low-lying linear excited state, a number of possibilities exist, most of which involve some degree of rehybridization. Similar arguments apply to the other members of the group, and all the possible transitions are tabulated in Table 10.4.

There are, as yet, no experimental data with which to compare these predictions, but the relatively large values predicted suggest that this group is a profitable field of study.

Table 10.4   The Electron Affinities of Group III

| Radical | Transition | Electron Affinity (kcal mole$^{-1}$) | | | |
|---------|-----------|:---:|:---:|:---:|:---:|
| | | B | Al | Ga | In |
| MH | $s^2p \rightarrow s^2pp$ | 2 | 1 | — | 2 |
| | $di^3di \rightarrow di^2di\pi$ | 17 | 22 | 28 | 21 |
| M$_2$H or MO | $didi\pi \rightarrow di^2di\pi$ | 49 | 60 | 72 | 64 |
| MH$_2$ | $trtrtr \rightarrow tr^2trtr$ | 32 | 49 | 53 | 44 |
| | $tetetete$ | 27 | 40 | 23 | 29 |
| | $trtrtr\pi$ | 24 | 38 | 24 | 28 |
| | $didi\pi\pi$ | 17 | 33 | 25 | 25 |
| | $didi\pi^2$ | 3 | 29 | 21 | 6 |
| | $te^2tete$ | — | 19 | 46 | 8 |
| | $trtr\pi^2$ | — | — | 23 | — |
| MH$_2$ | $didi\pi \rightarrow tr^2trtr$ | 39 | 53 | 57 | 47 |
| | $tetetete$ | 34 | 44 | 27 | 32 |
| | $trtrtr\pi$ | 31 | 42 | 28 | 32 |
| | $didi\pi\pi$ | 24 | 37 | 29 | 28 |
| | $didi\pi^2$ | 10 | 33 | 25 | 9 |
| | $te^2tete$ | — | 23 | 50 | 11 |
| | $trtr\pi^2$ | — | — | 27 | — |

## 10.9   The Predicted Electron Affinities of Group IVa

This group, embracing as it does all the organic ions which have been studied, not only offers many possibilities for the combination of valence states but also provides a great deal of experimental data wherewith to test the various predictions.

The possibility of triplet (divalent) states, brought into prominence by the current interest in the chemistry of the carbenes, must not be discounted in discussing possible valence states. This possibility brings into consideration not only carbene itself, CH$_2$, but also CH and similar radicals, despite the invalidity of the arguments of Hinze and Jaffe (the ionization of CH to CH$^-$ ($s^2pp \rightarrow s^2ppp$) does not correspond to the formation of a lone pair). The valence states of the carbenes require careful consideration. The lowest appropriate state of carbon is $s^2pp$.

corresponding to a bent, singlet state[8] for the radical, but the ground state is believed to be linear and triplet[8], corresponding to the valence state *didiππ* even though this lies 160 kcal above the $s^2pp$ state. The extra energy of the valence state permits a number of profitable rehybridizations in the ion. This linear ground state is not always observed in the carbenes, the radical $CF_2$[18] having the expected, bent, configuration.

Table 10.5   The Electron Affinities of Group IVa

| Radical | Transition | Electron Affinity (kcal mole$^{-1}$) | | | |
| | | C | Si | Ge | Sn |
|---|---|---|---|---|---|
| CH | $s^2pp$ → $s^2ppp$ | 16 | 28 | 18 | 39 |
| CH$_2$ | $s^2pp$ → $s^2ppp$ | 16 | 28 | 18 | 39 |
| CH$_2$ | *didiππ* → *didiπ$^2$π* | 2 | 80 | 95 | 115 |
| | *trtrtrπ$^2$* | 55 | 80 | 114 | 133 |
| | *te$^2$tetete* | 46 | 80 | 110 | 129 |
| CH$_3$ | *trtrtrπ* → *trtrtrπ$^2$* | 1 | 42 | 108 | 126 |
| | *te$^2$tetete* | 36 | 63 | 95 | 113 |
| CH$_3$ | *tetetete* → *trtrtrπ$^2$* | — | 39 | 107 | 124 |
| | *te$^2$tetete* | 31 | 60 | 94 | 111 |
| C$_6$H$_5$ or C$_2$H$_3$ | *trtrtrπ* → *tr$^2$trtrπ* | 45 | 73 | 113 | 131 |
| C$_2$H | *didiππ* → *di$^2$diππ* | 77 | 94 | 123 | 141 |

The possible combinations of levels may readily be deduced for carbon from the correlation diagram (Figure 10.1) even though no allowance has been made in that figure for the possibility of rehybridization. The heavier elements in this group show markedly increased electron affinities, in contrast to the earlier groups considered, and to the experimentally known electron affinities of the halogens at the other end of the periodic table. It must be remembered, however, that the interpolation of valence state energies must be held to be less accurate for these heavier elements, also the ground state electron affinities are estimates. Nevertheless, it would appear probable that there should be a very significant increase in the stabilities of the negative ions of these elements. It is unlikely that the multi-bonded atoms of any heavier element than carbon will be examined experimentally, although the possibility of mixed unsaturates remains, but the high predicted stabilities suggest that their existence as anionic intermediates must be taken seriously.

A considerable amount of experimental data has been amassed for the ions of this group, which is given in detail in Tables 11.1 to 11.6 and 12.3. The main results are summarized below in Table 10.6.

Table 10.6    Comparison of Observed and Predicted Electron Affinities

| Radical | Prediction | Observation | Reference |
|---------|-----------|-------------|-----------|
| CH | 16 | 21 | — |
| $CH_3$ | 31 to 36 | 20 to 24 | Table 11.2 |
| $C_6H_5$ | 45 | 44 to 66 | Table 11.6 |
| $C_2H$ | 77 | 62 | Table 11.5 |
| $CF_3$ | 31 to 36 | 42 | Table 11.7 |
| $SiF_3$ | 60 to 63 | 77 | Table 12.3 |

## 10.10    The Electron Affinities of Group V

This group presents a number of problems in the prediction of electron affinities, because of the possibility of lone pair hybridization. Such hybridization will inevitably increase the electron affinity by increasing the $s$ character of the orbital accepting the electron, but the amount of lone pair hybridization may be slight, as suggested by Peters[19] in discussing the ionization potential of ammonia, rising to complete hybridization in $NO_2$. It is not clear if the high electron affinity in $NF_2$ is to be attributed to such hybridization, accompanied by a marked shortening in the NF bond, or to the breakdown of the basic assumption about the bond energy in the ion, as occurs in $CF_3$. The increase in electron affinity of $NF_2$ over that

Table 10.7

| Radical | Transition | Electron Affinity (kcal mole$^{-1}$) | | | |
|---------|-----------|---|---|----|----|
| | | N | P | As | Sb |
| $MH_2$ | $s^2ppp \rightarrow s^2p^2pp$ | 18 | 28 | 31 | 28 |
| | $tr^2trtr\pi \rightarrow tr^2trtr\pi^2$ | 40 | 41 | 61 | — |
| | $trtrtr\pi^2 \rightarrow tr^2trtr\pi^2$ | 113 | 86 | 104 | 10 |
| | $te^2tetete \rightarrow te^2te^2tete$ | 95 | 74 | 64 | 10 |
| MCX $\rbrace$ MMM $\rbrace$ | $didi\pi\pi \rightarrow di^2di\pi\pi$ | 157 | 110 | 122 | 23 |
| | $tr^2trtr\pi \rightarrow tr^2tr^2tr\pi$ | 117 | 86 | 93 | 20 |
| $MH_2$ | $didi\pi^2\pi \rightarrow didi\pi^2\pi^2$ | 48 | 29 | 60 | — |

of $NH_2$ is much greater than the corresponding increase in $CF_3$, which argues that a change in hybridization must also contribute.

The only structure for the $MH_2$ radical which leads to reasonable prediction is the $s^2ppp$ state. It is doubtful if the valence state can be expressed as simply as this and, as already pointed out, some degree of lone pair hybridization may be expected, and indeed must occur with $NO_2$ (E = 92 kcal). The electron affinities found for $NH_2$ and doubly substituted $NH_2$ radicals are uniformly 26 kcal mole$^{-1}$ (35 kcal mole$^{-1}$ if singly substituted) and for $PH_2$ 37 kcal mole$^{-1}$, the difference reflecting the difference in the predicted affinities.

The chalcocyanates are undoubtedly derived from the trigonal HNCX, the structure at least of thiocyanic acid being established by microwave spectroscopy, but the observed affinities suggest that the ions may more properly be referred to sulphur or selenium. Thermochemically this will not affect the observations.

### 10.11 The Electron Affinities of Groups IV and VII

There are few possibilities of variation in the electron affinities of these groups. Group VII, the halogens, is confined to observation of the atomic ions, while only the single state of the atom is of importance in Group IV, all hybridized states involving much higher promotion energies.

Table 10.8

| Radical | Transition | Electron Affinity (kcal mole$^{-1}$) | | | |
|---------|-----------|-----|-----|-----|-----|
| | | O | S | Se | Te |
| MH | $s^2p^1pp \rightarrow s^2p^2p^2p$ | 46 | 55 | 58 | 59 |

The electron affinity of OH is now accepted as 42 kcal mole$^{-1}$ [20], a good agreement with the predicted value. Some doubt exists about the affinities of the alkoxide ions, Hamill[17] placing them above hydroxyl, but several studies in the magnetron place them below hydroxyl, whether they are derived from the alcohol or the peroxide. The magnetron result for SH is in good agreement with the values determined in flames[21], by mass spectrometry[22], or by photoionization[23], and accords closely with the predicted value, but the affinities of the mercaptyl radicals are lower by 20 kcal mole$^{-1}$.

The valence state of sulphur in $SF_5$ is not included in the tables of Hinze, so that no comparison with prediction is possible. The suggestion that the chalcocyanates should be included in this group would place them close to the predicted values (SCN; 50 kcal mole$^{-1}$ [24], SeCN 62 kcal mole$^{-1}$ [25].

5+

**REFERENCES**

1. Cottrell, T. L., *The Strengths of Chemical Bonds*, Butterworths, London, 1958.
2. Hinze, J. A., *Ph.D. Thesis*, University of Cincinnati, 1962.
3. Hinze, J. A. and Jaffe, H. H., *J. Amer. Chem. Soc.*, **84**, 540 (1962).
4. Hinze, J. A. and Jaffe, H. H., *U.S. Air Force Rept.*, 1963.
5. Mulliken, R. S., *Tetrahedron*, **6**, 68 (1959).
6. Rohrlich, F., *Phys. Rev.*, **69**, 101 (1956).
7. Edlen, B., *J. Chem. Phys.*, **33**, 98 (1960).
8. Ginsberg, A. P. and Miller, J. M., *J. Inorg. Nucl. Chem.*, **7**, 351 (1958).
9. Branscomb, L. M. and Smith, S. J., *J. Chem. Phys.*, **25**, 598 (1956).
10. Branscomb, L. M., Burch, D. S., Smith, S. J. and Geltman, S., *Phys. Rev.*, **111**, 504 (1958).
11. Doerffel, K., *Z. Anorg. Allg. Chem.*, **281**, 212 (1955).
12. Cubicciotti, D. D., *J. Chem. Phys.*, **31**, 1646 (1959).
13. Bailey, T. L., *J. Chem. Phys.*, **28**, 792 (1958).
14. Branscomb, L. M., Steiner, B. and Seman, M. L., *Atomic Collision Processes*, North Holland, Amsterdam, 1964.
15. Berry, R. S. and Reimann, C. W., *J. Chem. Phys.*, **38**, 1540 (1963).
16. Gaines, A. F. and Page, F. M., *Trans. Faraday Soc.*, **62**, 3086 (1966).
17. Hamill, W. H., *Private communication*.
18. Ramsay, D. A., *Ann. N.Y. Acad. Sci.*, **67**, 485 (1957).
19. Peters, D., *J. Chem. Soc.*, **3**, 2901 (1964).
20. Branscomb, L. M., *J. I. L. A. Rept.*, No. 62, University of Colorado, 1966.
21. Page, F. M., *Rev. Inst. Français Petrole*, **13**, 692 (1958).
22. Steiner, B., *Private communication*.
23. Chupka, W. A., *Private communication*.
24. Napper, R. and Page, F. M., *Trans. Faraday Soc.*, **59**, 1086 (1963).
25. Newman, R. N., *B.Sc. Thesis*, The University of Aston in Birmingham, 1967.
26. Herzberg, G., *Spectra of Diatomic Molecules*, Van Nostrand, New York, 1940.

# 11

# σ-Capture by Carbon

## 11.1 Introduction

A very large number of the negative ions which have been observed are believed to be closed shell (singlet) ions in which the extra electron forms a lone pair, and might be regarded as a normal substituent. The strongest evidence that the electron is, indeed, localized as a lone pair on one atom, the acceptor atom, is the insensitivity of the electron affinity to structural changes, which is clearly shown in the tables in this and the next chapter.

## 11.2 Valence States of Carbon Anions

Carbon may exist in three states of hybridization: (1) acetylenic, digonal or $sp$; (2) ethylenic, trigonal or $sp^2$ and (3) alkyl, tetrahedral or $sp^3$. It has been shown that the electron affinities to be expected of carbon atoms in these states of hybridization are 77, 45 and 31 kcal mole$^{-1}$ [1] respectively, and these theoretical estimates are borne out in the tables below. These tables have been arranged according to the mechanism of ion formation, types II, III and IV are subdivided according to the state of hybridization. Ions formed by type III mechanism require a knowledge of a bond dissociation energy $D$, and those formed by type IV mechanism require a knowledge also of the heat of adsorption of any residue remaining on the filament.

Bond dissociation energies, unless otherwise indicated, are taken from *The Strengths of Chemical Bonds*, by T. L. Cottrell.

117

## 11.3 Type 2 Split

This involves rupture of the bond in the gaseous phase and only molecules containing a weak bond will exhibit this behaviour. Hence $E' = E$.

*Digonal Ions.* No examples found.

Table 11.1    Trigonal Ions

| Compound | Formula | $E_T$ | $T°K$ | Ion | Filament | $E_0$ | |
|---|---|---|---|---|---|---|---|
| 2:2'-dichlorobenzil | $C_{14}Cl_2H_8O_2$ | 61·5 ± 2·5 | 1410 | $C_6ClH_4{}^-$ | W | 55·9 ± 2·5 | Note 1 |
| 3:3'-dichlorobenzil | $C_{14}Cl_2H_8O_2$ | 58·9 ± 2·9 | 1410 | $C_6ClH_4{}^-$ | W | 53·3 ± 2·9 | Note 1 |
| Benzil | $C_{14}H_{10}O_2$ | 55·6 ± 1·0 | 1560 | $C_6H_5{}^-$ | W | 49·4 ± 1·0 | Note 1 |
| 3:3'-dinitrobenzil | $C_{14}H_{12}N_2O_4$ | 44·9 ± 3·7 | 1425 | $C_6H_4NO_2{}^-$ | Ir | 39·2 ± 3·7 | Note 1 |

*Note 1*: Benzil and its derivatives are believed to produce phenyl ions by a symmetrical fission, followed rapidly by the elimination of carbon monoxide. In the case of dimethoxybenzil the elimination of carbon dioxide appears to be a slow process and the results cannot be similarly interpreted.

Table 11.2 Tetrahedral Ions

| Compound | Formula | $E_T$ | $T°K$ | Ion | Filament | $E_0$ | |
|---|---|---|---|---|---|---|---|
| Fluoroform | $CF_3H$ | $48·3 ± 5·2$ | 1562 | $CF_3{}^-$ | Pt | $42·0 ± 5·2$ | Note 1 |
| Mercury dimethyl | $C_2H_6Hg$ | 31·1 | 1700 | $CH_3{}^-$ | W | 24·3 | |
| Mercury diethyl | $C_4H_{10}Hg$ | 27·1 | 1700 | $C_2H_5{}^-$ | W | 20·3 | |
| Lead tetramethyl | $C_4H_{12}Pb$ | 31·1 | 1700 | $CH_3{}^-$ | W | 24·3 | |
| Mercury di-*n*-propyl | $C_6H_{14}Hg$ | 21·6 | 1750 | $C_3H_7{}^-$ | W | 14·6 | |
| Mercury di-*n*-butyl | $C_8H_{18}Hg$ | 20·6 | 1750 | $C_4H_9{}^-$ | W | 13·6 | |
| Lead tetraethyl | $C_8H_{20}Pb$ | 27·3 | 1640 | $C_2H_5{}^-$ | W | 20·7 | |
| Dibenzyl | $C_{14}H_{14}$ | 23·9 | 1625 | $C_6H_5CH_2{}^-$ | Ir | 17·4 | Note 2 |
| Dibenzyl | $C_{14}H_{14}$ | 25·5 | 1568 | $C_6H_5CH_2{}^-$ | W | 19·2 | Note 2 |

*Note 1:* Energetically, this result corresponds closely to the electron affinity of $CF_3$, as determined from other substances, and since it occurs at a higher temperature than the corresponding type IV mechanism, it is included here.

*Note 2:* These results were obtained in the presence of silver surfaces. In a vessel without such surfaces, the mechanism was type 3 involving the C—C bond energy.

5*

## 11.4  Type 3 Split

In this case the parent molecule has a high bond energy so only a small fraction is dissociated. Fission occurs on the hot filament to give electron acceptors. Hence $E' = E - D$.

Table 11.3  Digonal Ions

| Compound | Formula | $E_T'$ | $T°K$ | $E_0'$ | Ion | Filament | $D$ | $E_0$ | |
|---|---|---|---|---|---|---|---|---|---|
| Cyanogen | $C_2N_2$ | $16.0 \pm 1.1$ | 1630 | $9.5 \pm 1.1$ | $CN^-$ | W | 130 | $74.5 \pm 1.1$ | Note 1 |
| Diacetylene | $C_4H_2$ | 3.6 | 1410 | $-2.0$ | $C_2H^-$ | Pt | 126 | 61.0 | |

*Note 1*: The precise value for the electron affinity depends upon that assumed for the bond dissociation energy. Using Dibeler's value for $\Delta H_f°(CN)$ of 101.5 kcal mole$^{-1\,2}$ gives $D(NC-CN) = 130$ kcal mole$^{-1}$.

Table 11.4  Trigonal Ions

| Compound | Formula | $E_T'$ | $T°K$ | $E_0'$ | Ion | Filament | $D$ | $E_0$ | |
|---|---|---|---|---|---|---|---|---|---|
| p-Dibromobenzene | $Br_2C_6H_4$ | $-5.3$ | 1430 | $-11.0$ | $BrC_6H_4$ | Ir | 72 | 61.0 | Note 1 |
| Tetrafluoroethylene | $C_2F_4$ | 10.6 | 1400 | 5.0 | $CF_2^-$ | Pt | 112 | 61.0 | Note 2 |
| Azulene | $C_{10}H_8$ | $-34.3$ | 1600 | $-40.7$ | $C_{10}H_7^-$ | W | 102 | 61.3 | Note 3 |

*Note 1*: See also Table 11.7.
*Note 2*: The valence state involved in this capture is uncertain; the ion could be in a tetrahedral state.
*Note 3*: The C—H bond dissociation energy used is that quoted for benzene.

Table 11.5  Tetrahedral Ions

| Compound | Formula | $E_T'$ | $T°K$ | $E_0'$ | Ion | Fil. | $D$ | $E_0$ |
|---|---|---|---|---|---|---|---|---|
| Bromoform | $Br_3CH$ | 0·1 ± 0·2 | >1350 | −5·3 ± 0·2 | $Br_2CH^-$ | Pt | 49 | 44·7 ± 0·2 |
| Carbon tetra-bromide | $Br_4C$ | 0·4 ± 1·0 | 1550 | −5·8 ± 1·0 | $Br_3C^-$ | Pt | 49 | 43·2 ± 1·0 |
| Fluoroform | $CF_3H$ | −46·1 ± 3·2 | 1471 | −52·0 ± 3·2 | $CF_3^-$ | Pt | 103 | 51·0 ± 3·2 |
| Hexafluoroethane | $C_2F_6$ | 3·9 ± 2·4 | 1520 | −2·2 ± 2·4 | $CF_3^-$ | Ir | 97 | 46·3 ± 2·4 |
| Lead tetraethyl | $C_8H_{20}Pb$ ($C_2H_5$) | −48 | 1880 | −55·5 | $CH_3^-$ | W | $D(CH_3{-}CH_2)$ =79·7 | 24·2  Note 1 |

*Note 1:* The observed result was clearly defined, but the interpretation is open to doubt.

## 11.5  Type 4 Split

This involves heterolytic fission to give one electron acceptor, with adsorption of the residue on the filament. Hence $E' = E + Q - D$.

Table 11.6  Digonal Ions

| Compound | Formula | $E_T'$ | $T°K$ | $E_0'$ | Ion | Fil. | Residue | $Q$ | $D$ | $E_0$ |
|---|---|---|---|---|---|---|---|---|---|---|
| Hydrogen cyanide | CHN | 29·8 ± 1·7 | 1870 | 18·6 ± 1·7 | $CN^-$ | W | H | 72 | 124 | 70·6 ± 1·7  Note 1 |
| Acetylene | $C_2H_2$ | 29·9 ± 3·9 | 1620 | 20·2 ± 3·9 | $C_2H^-$ | WC | H | 72 | 114 | 62·2 ± 4·6  Note 2 |
| Phenylacetylene | $C_8H_6$ | 33·5 | 1535 | 24·3 | $C_8H_5^-$ | WC | H | 72 | 114 | 66·3 |

*Note 1:* D(H—CN) energy is derived using Dibeler's value for $\Delta H_f°(CN)$[1].
*Note 2:* The C—H bond dissociation energy is that quoted for acetylene. An alternative mechanism would be the formation of a substituted phenyl ion with an electron affinity of 54·3 kcal mole.

Table 11.7 Trigonal Ions

| Compound | Formula | $E_T'$ | $T°K$ | $E_0'$ | Ion | Fil. | Res. | $Q$ | $D$ | $E_0$ | |
|---|---|---|---|---|---|---|---|---|---|---|---|
| Pentafluorobromobenzene | $BrC_6F_5$ | $17·2 ± 0·1$ | 1220 | $9·9 ± 0·1$ | $C_6F_5^-$ | Ir | Br | 17 | 71 | $63·9 ± 0·1$ | |
| Pentafluorobromobenzene | $BrC_6F_5$ | 0 | 1400 | $-8·3$ | $BrC_6F_4^-$ | Ir | F | 48·8 | 121 | 63·3 | Note 1 |
| p-Dibromobenzene | $Br_2C_6H_4$ | 14·2 | 1500 | 5·3 | $BrC_6H_4^-$ | W | Br | 17 | 71 | 59·3 | Note 2 |
| Allene | $C_3H_4$ | 27·0 | 1537 | 17·8 | $C_3H_3^-$ | Ir | H | 66 | 102 | 53·8 | Note 3 |
| Octafluorobut-2-ene | $C_4F_8$ | $-0·6 ± 1·1$ | 1600 | $-9·3 ± 1·1$ | $C_4F_7^-$ | Pt | F | 51·8 | 123 | $61·9 ± 1·1$ | |
| Fumaronitrile | $C_4H_2N_2$ | $18·7 ± 1·0$ | 1430 | $10·1 ± 1·0$ | $C_4HN_2^-$ | WC | H | 72 | 102 | $40·1 ± 1·0$ | Note 3 |
| Hexachlorobenzene | $C_6Cl_6$ | 17·1 | 1500 | $8·1 ± 2·3$ | $C_6Cl_5^-$ | Ir | Cl | 35·0 | 90·5 | $63·6 ± 2·3$ | |
| Pentafluorochlorobenzene | $C_6ClF_5$ | $17·5 ± 0·5$ | 1390 | 9·2 | $C_6F_5^-$ | Ir | Cl | 35 | 90·5 | 64·7 | Note 4 |
| 1-Chloro-2-nitrobenzene | $C_6ClH_4NO_2$ | $14·5 ± 0·8$ | 1500 | $5·6 ± 0·8$ | $C_6H_4NO_2^-$ | W | Cl | 42 | 90·5 | $54·1 ± 0·8$ | Note 4 |
| 1-Chloro-3-nitrobenzene | $C_6ClH_4NO_2$ | $-9·5 ± 0·3$ | 1500 | $-18·4 ± 0·3$ | $C_6H_4NO_2^-$ | Ir | Cl | 35 | 90·5 | $37·1 ± 0·3$ | Note 4 |
| 1-Chloro-3-nitrobenzene | $C_6ClH_4NO_2$ | 0 | 1500 | $-8·9$ | $C_6H_4NO_2^-$ | W | Cl | 42 | 90·5 | 38·6 | Note 4 |
| 1-Chloro-4-nitrobenzene | $C_6ClH_4NO_2$ | $16·2 ± 1·9$ | 1450 | $7·6 ± 1·9$ | $C_6H_4NO_2^-$ | W | Cl | 42 | 90·5 | $56·1 ± 1·9$ | Note 4 |
| Chlorobenzene | $C_6ClH_5$ | $9·4 ± 1·6$ | 1410 | $0·9 ± 1·6$ | $C_6H_5^-$ | W | Cl | 42 | 90·5 | 49·4 | Note 5 |
| 1:2-Dichlorobenzene | $C_6Cl_2H_4$ | $5·7 ± 0·7$ | 1490 | $-3·2 ± 0·7$ | $C_6ClH_4^-$ | Ir | Cl | 35 | 90·5 | $52·3 ± 0·7$ | Note 4 |
| 1:2-Dichlorobenzene | $C_6Cl_2H_4$ | $13·3 ± 1·6$ | 1400 | $4·9 ± 1·6$ | $C_6ClH_4^-$ | W | Cl | 42 | 90·5 | $53·4 ± 1·6$ | Note 4 |
| 1:3-Dichlorobenzene | $C_6Cl_2H_4$ | $7·8 ± 1·4$ | 1410 | $-0·6 ± 1·4$ | $C_6ClH_4^-$ | W | Cl | 42 | 90·5 | $47·9 ± 1·4$ | Note 4 |
| 1:3-Dichlorobenzene | $C_6Cl_2H_4$ | $0·5 ± 2·0$ | 1490 | $-8·4 ± 2·0$ | $C_6ClH_4^-$ | Ir | Cl | 35 | 90·5 | $46·9 ± 2·0$ | Note 4 |
| 1:4-Dichlorobenzene | $C_6Cl_2H_4$ | $13·3 ± 1·8$ | 1410 | $4·9 ± 1·8$ | $C_6ClH_4^-$ | W | Cl | 42 | 90·5 | $53·4 ± 1·8$ | Note 4 |

| Compound | Formula | | | | Ion | | | | | | Note |
|---|---|---|---|---|---|---|---|---|---|---|---|
| 1:4-Dichlorobenzene | $C_6Cl_2H_4$ | 8·3 ± 1·8 | 1500 | −0·6 ± 1·8 | $C_6ClH_4^-$ | Ir | Cl | 35 | 90·5 | 54·1 ± 1·8 | Note 4 |
| 1:4-Dichlorobenzene | $C_6Cl_2H_4$ | 26·1 | 1500 | 17·2 | $C_6Cl_2H_3^-$ | W | H | 72 | 102 | 47·2 | Note 3 |
| 1:2:3-Trichlorobenzene | $C_6Cl_3H_3$ | 21·5 ± 1·4 | 1410 | 13·1 ± 1·4 | $C_6Cl_2H_3^-$ | W | Cl | 42 | 90·5 | 61·6 ± 1·4 | Note 4 |
| 1:3:5-Trichlorobenzene | $C_6Cl_3H_3$ | 11·5 ± 1·0 | 1410 | 3·1 ± 1·0 | $C_6Cl_2H_3^-$ | W | Cl | 42 | 90·5 | 51·6 ± 1·0 | Note 4 |
| 1:2:3:4-Tetrachlorobenzene | $C_6Cl_4H_2$ | 15·8 ± 1·3 | 1410 | 7·4 ± 1·3 | $C_6Cl_3H_2^-$ | W | Cl | 42 | 90·5 | 55·9 ± 1·3 | Note 4 |
| 1:2:4:5-Tetrachlorobenzene | $C_6Cl_4H_2$ | 18·8 ± 2·2 | 1410 | 10·4 ± 2·2 | $C_6Cl_3H_2^-$ | W | Cl | 42 | 90·5 | 58·9 ± 2·2 | Note 4 |
| Fluorobenzoquinone | $C_6FH_3O_2$ | 28·6 ± 2·8 | 1590 | 19·1 ± 2·8 | $C_6FH_2O_2^-$ | Ir | H | 66 | 102 | 55·1 ± 2·8 | Note 3 |
| Hexafluorobenzene | $C_6F_6$ | 0 | 1500 | −8·9 | $C_6F_5^-$ | Ir | F | 48·8 | 121 | 63·3 | Note 1 |
| p-Benzoquinone | $C_6H_4O_2$ | 24·8 | 1470 | 16·0 | $C_6H_3O_2^-$ | WC | H | 72 | 102 | 46·0 | Note 3 |
| Benzene | $C_6H_6$ | 28·5 ± 1·4 | 1300 | 20·7 ± 1·4 | $C_6H_5^-$ | Ir | H | 66 | 102 | 56·7 ± 1·4 | Note 3 |
| Benzene | $C_6H_6$ | 34·5 ± 1·2 | 1560 | 25·1 ± 1·2 | $C_6H_5^-$ | W | H | 72 | 102 | 55·1 ± 1·2 | Note 3 |
| 2:3-Dicyanobenzoquinone | $C_8H_2N_2O_2$ | 14·8 ± 1·8 | 1490 | 5·9 ± 1·8 | $C_8HN_2O_2^-$ | Ir | H | 66 | 102 | 41·9 ± 1·8 | Note 3 |
| Tetrafluoronaphthalene | $C_{10}F_4H_4$ | 22·8 | 1510 | 13·8 | $C_{10}F_4H_3^-$ | Ir | H | 66 | 102 | 49·8 | Note 3 |
| Octafluoronaphthalene | $C_{10}F_8$ | −5·0 | 1475 | −13·8 | $C_{10}F_7^-$ | Ir | F | 49 | 121 | 58·2 | Note 1 |
| Naphthoquinone | $C_{10}H_6O_2$ | 23·1 ± 1·4 | 1405 | 14·7 ± 1·4 | $C_{10}H_5O_2^-$ | Ir | H | 66 | 102 | 50·7 ± 1·4 | Note 3 |
| Naphthalene | $C_{10}H_8$ | 23·0 ± 0·5 | 1330 | 15·0 ± 0·5 | $C_{10}H_7^-$ | Ir | H | 66 | 102 | 51·0 ± 0·5 | Note 3 |
| Mercury diphenyl | $C_{12}H_{10}Hg$ | 33·1 ± 1·8 | 1500 | 24·2 ± 1·8 | $C_6H_5^-$ | W | Hg | 2 × 57 | 86 | 52·2 ± 1·8 | Note 6 |
| Anthracene | $C_{14}H_{10}$ | 20·9 ± 0·4 | 1400 | 12·6 ± 0·4 | $C_{14}H_9^-$ | Ir | H | 66 | 102 | 48·6 ± 0·4 | Note 3 |
| Mercury di-p-tolyl | $C_{14}H_{14}Hg$ | 38·6 ± 2·8 | 1423 | 30·1 ± 2·8 | $CH_3C_6H_4^-$ | W | Hg | 2 × 57 | 86 | 58·1 ± 2·8 | Note 6 |

Note 1: Most perfluoroaromatic systems showed a region of thermoneutral ion formation which is ascribed to a similar process in each case.

Note 2: See also Table 11.4.

Note 3: The C—H bond dissociation energy used is that quoted for benzene.

Note 4: The C—Cl bond dissociation energy used is that derived for chlorobenzene[3].

Note 5: This compound was used to establish the D(C—Cl) energy in the aromatic series by comparison with the phenyl ion.

Note 6: Q is here not a heat absorption, but is an excitation energy evolved when the mercury atom reverts to its thermochemical ground state.

Table 11.8  Tetrahedral Ions

| Compound | Formula | $E_T'$ | $T°K$ | $E_0'$ | Ion | Fil. | Res. | Q | D | $E_0$ | |
|---|---|---|---|---|---|---|---|---|---|---|---|
| Monobromotri-fluoromethane | $BrCF_3$ | $0.5 \pm 1.9$ | 1408 | $-8.0 \pm 1.9$ | $CF_3^-$ | Pt | Br | 19 | 68 | $41.0 \pm 1.9$ | |
| Bromoform | $Br_3CH$ | $23.0 \pm 6.0$ | 1285 | $15.4 \pm 6.0$ | $CBr_3^-$ | Pt | H | 69 | 95 | $41.4 \pm 6.0$ | |
| Carbon tetra-bromide | $Br_4C$ | $19.4 \pm 2.0$ | 1365 | $11.3 \pm 2.0$ | $CBr_3^-$ | Pt | Br | 25 | 49 | $35.3 \pm 2.0$ | Note 1 |
| Monochlorotri-fluoromethane | $CClF_3$ | $0.5 \pm 1.4$ | 1724 | $-9.5 \pm 1.4$ | $CF_3^-$ | Ir | Cl | 33.5 | 83 | $40.0 \pm 1.4$ | |
| Chloroform | $CCl_3H$ | $15.7 \pm 0.7$ | 1450 | $7.0 \pm 0.7$ | $CCl_3^-$ | Pt | H | 69 | 90 | $28.0 \pm 0.7$ | |
| Carbon tetra-chloride | $CCl_4$ | $7.3 \pm 0.9$ | 1430 | $-1.2 \pm 0.9$ | $CCl_3^-$ | Pt | Cl | 40.7  1.9 | 69.9 | $28.0 \pm 0.7$ | Note 1 |
| Fluoroform | $CF_3H$ | $16.7 \pm 1.8$ | 1342 | $8.7 \pm 1.8$ | $CF_3^-$ | Pt | H | 69 | 103 | $42.7 \pm 1.8$ | |
| Carbon tetra-fluoride | $CF_4$ | $-18.2 \pm 1.1$ | 1390 | $-26.5 \pm 1.1$ | $CF_3^-$ | Pt | F | 51.8  2.8 | 121 | $42.7 \pm 1.8$ | Note 1 |
| Hexachloroethane | $C_2Cl_6$ | $9.9 \pm 1.8$ | 1470 | $1.2 \pm 1.8$ | $C_2Cl_5^-$ | Pt | Cl | 35.3  1.5 | 69.9 | $35.8 \pm 3.3$ | |
| Octafluoro-propane | $C_3F_8$ | $-15.1 \pm 4.2$ | 1600 | $-24.5 \pm 4.2$ | $C_3F_7^-$ | Pt | F | 51.8 | 122.2 | $45.9 \pm 4.2$ | |
| Toluene | $C_7H_8$ | $22.1 \pm 1.9$ | 1540 | $12.9 \pm 1.9$ | $C_6H_5CH_2^-$ | W | H | 72 | 78 | $18.9 \pm 1.9$ | |
| Duroquinone | $C_{10}H_{12}O_2$ | $15.1 \pm 1.1$ | 1320 | $7.2 \pm 1.1$ | $C_{10}H_{11}O_2^-$ | Ir | H | 66 | 78 | $18.5 \pm 1.1$ | Note 2 |
| Duroquinone | $C_{10}H_{12}O_2$ | $21.5 \pm 2.7$ | 1600 | $12.0 \pm 2.7$ | $C_{10}H_{11}O_2^-$ | WC | H | 72 | 78 | $18.0 \pm 2.7$ | Note 2 |

*Note 1*: The heats of adsorption of the halogen atoms on platinum were determined from these compounds.
*Note 2*: D(C—H) used is that quoted for toluene.

## REFERENCES

1. Gaines, A. F. and Page, F. M., *Trans. Faraday Soc.*, **62**, 3086 (1966).
2. Dibeler, V. H. and Liston, S. K., *J. Chem. Phys.*, **47**, 4548 (1967).
3. Moss, D., *M.Sc. Thesis*, University of Wales, 1967.

# σ-Capture by Other Elements

## 12.1 Introduction

The general levels of the electron affinities of elements other than carbon may be predicted from the promotion energies of the element to the valence state, and the electron affinity in the ground state. Extensive tables of these promotion energies are available, and predictions of the electron affinities of most elements in the first two rows of the periodic table have been made. These predictions are, however, incomplete, since promotion energies to certain valence states are not available, nor have all valence states for which predictions are available been observed experimentally. The following electron affinities are among those predicted:

| $CH_3$ | $NH_2$ | OH | F | |
|--------|--------|-----|------|------------------------|
| 31 | 19 | 46 | (80) | kcal mole$^{-1}$ |
| $SiH_3$ | $PH_2$ | SH | Cl | |
| 63 | 37 | 55 | (84) | kcal mole$^{-1}$ |

## 12.2 Experimental Results

The experimental data collected in the following tables agree quite closely with these predictions. Even though only $SiF_3^-$ has been studied, the difference between its stability (77 kcal mole$^{-1}$) and that of $CF_3^-$ (42 kcal mole$^{-1}$) is close to the predicted difference between the electron affinities of the corresponding hydrides $SiH_3$ and $CH_3$.

126

## 6

Table 12.1  σ-Nitrogen Ions

| Compound | Formula | $E_T'$ | $T°K$ | $E_0'$ | Type of split | Ion | Fil. | Res. | Q | D | $E_0$ | |
|---|---|---|---|---|---|---|---|---|---|---|---|---|
| Thiocyanic acid | CHNS | 17·0 ± 0·6 | 1870 | 5·8 ± 0·6 | IV | $SCN^-$ | W | H | 72 | 114·6 ± 1·1 | 48·4 ± 0·5 | Note 1 |
| Dimethylamine | $C_2H_7N$ | 9·0 | 1500 | 0·0 | IV | $C_2H_6N^-$ | W | H | 72 | 96·0 | 24·0 | Note 3 |
| Thiocyanogen | $C_2N_2S_2$ | 55·6 ± 0·5 | 1800 | 48·4 ± 0·5 | II | $SCN^-$ | W | — | — | — | 48·4 ± 0·5 | Note 1 |
| Selenocyanogen | $C_2N_2Se$ | 69·9 | 1500 | 63·7 | II | $SeCN^-$ | W | — | — | — | 63·7 | Note 1 |
| Tetramethyl-tetrazene | $C_4H_{12}N_4$ | 29·6 | 1400 | 24·0 | II | $C_2H_6N^-$ | Pt | — | — | — | 24·0 | |
| Pentafluoro-aniline | $C_6F_5H_2N$ | 12·8 ± 2·3 | 1450 | 4·1 ± 2·3 | IV | $C_6F_5NH^-$ | Ir | H | 66 | 100 | 38·1 ± 2·3 | Note 2 |
| Aniline | $C_6H_7N$ | 11·7 ± 2·5 | 1667 | 1·7 ± 2·5 | IV | $C_6H_6N^-$ | Ir | H | 66 | 100 | 35·7 ± 1·5 | Note 3 |
| Methylaniline | $C_7H_9N$ | 16·0 | 1500 | 7·0 | IV | $C_7H_8N^-$ | Ir | H | 66 | 89·0 | 30·0 | Note 3 |
| Diphenylamine | $C_{12}H_{11}N$ | 12·0 ± 1·8 | 1653 | 2·1 ± 1·8 | IV | $C_{12}H_{10}N^-$ | Ir | H | 66 | 91·2 ± 5·3 | 27·3 ± 3·5 | Note 3 |
| Diphenyl hydrazine | $C_{12}H_{12}N_2$ | 43·8 ± 1·5 | 2012 | 35·7 ± 1·5 | II | $C_6H_6N^-$ | W | — | — | — | 35·7 ± 1·5 | |
| 1:4-Dimethyl-1:4-diphenyl-tetrazene | $C_{14}H_{16}N_4$ | 35·6 | 1400 | 30·0 | II | $C_7H_8N^-$ | Pt | — | — | — | 30·0 | |
| Tetraphenyl hydrazine | $C_{24}H_{20}N_2$ | 33·2 ± 3·5 | 1980 | 27·3 ± 3·5 | II | $C_{12}H_{10}N^-$ | W | — | — | — | 27·3 ± 3·5 | |
| Tetrafluoro-hydrazine | $F_4N_2$ | 74·0 | 1250 | 69·0 | II | $NF_2^-$ | Pt | — | — | — | 69·0 | Note 4 |
| Ammonia | $H_3N$ | 3·0 | 1700 | -7·2 | IV | $NH_2^-$ | W | H | 72 | 104·9 ± 2·0 | 25·7 ± 2·0 | Note 3 |
| Hydrazine | $H_4N_2$ | 33·3 ± 2·0 | 1900 | 25·7 ± 2·0 | II | $NH_2^-$ | W | — | — | — | 25·7 ± 2·0 | |
| Nitric oxide | NO | 26·5 ± 2·5 | 1630 | 19·1 ± 2·5 | I | $NO^-$ | Pt | — | — | — | 19·1 ± 2·5 | |

(continued)

Table 12.1 (*continued*)

| Compound | Formula | $E_T'$ | $T°K$ | $E_0'$ | Type of Split | Ion | Fil. Res. | $Q$ | $D$ | $E_0$ |
|---|---|---|---|---|---|---|---|---|---|---|
| Nitrogen dioxide | $NO_2$ | $96.2 \pm 3.7$ | 1537 | $90.0 \pm 3.7$ | I | $NO_2^-$ | Pt | — | — | $90.0 \pm 3.7$ |
| Nitrogen dioxide | $NO_2*$ | $5.0 \pm 2.2$ | 1670 | $0.0 \pm 2.2$ | I | $NO_2^-*$ | Pt | — | — | $0.0 \pm 2.2$ |

*Note 1*: There is some doubt as to whether these results should be tabulated here. [The structure of thiocyanic acid is certainly HNCS and the N—H bond energy derived by the comparison of the apparent electron affinities is very reasonable but the magnitude of the electron affinity suggests $\sigma$-capture by the more electronegative sulphur, and treating the SCN ion as a localized lone pair ion may be an oversimplification.]

*Note 2*: $D(N—H)$ used is that derived for aniline.

*Note 3*: The $D(N—H)$ energies were derived for these compounds by comparison with the corresponding hydrazines or tetrazenes.

*Note 4*: Although only a preliminary result this shows very clearly the alteration of N hybridization by fluorine.

Table 12.2  σ-Oxygen Ions

| Compound | Formula | $E_T'$ | $T°K$ | $E_0'$ | Type of Split | Ion | Fil. | Res. | $Q$ | $D$ | $E$ | |
|---|---|---|---|---|---|---|---|---|---|---|---|---|
| Carbon dioxide | $CO_2$ | −90 | 1800 | −97·2 | III | $O^-$ | W | — | — | 129·0 | 31·8 | |
| Methanol | $CH_4O$ | −15·8 ± 0·4 | 1430 | −24·3 ± 0·4 | IV | $CH_3O^-$ | Pt | H | 69 | 102 | 8·7 ± 0·4 | |
| Bis(trifluoromethyl) peroxide | $C_2F_6O_2$ | 36·8 | 1400 | 31·2 | II | $CF_3O^-$ | Pt | — | — | — | 31·2 | |
| Ethanol | $C_2H_6O$ | −10·7 ± 1·1 | 1430 | −19·3 ± 1·1 | IV | $C_2H_5O^-$ | Pt | H | 69 | 102 | 13·7 ± 1·1 | |
| Iso propanol | $C_3H_8O$ | −9·0 ± 0·3 | 1430 | −17·5 ± 0·3 | IV | $C_3H_7O^-$ | Pt | H | 69 | 102 | 15·5 ± 0·3 | |
| n-butyl alcohol | $C_4H_{10}O$ | −7·8 ± 1·4 | 1720 | −18·1 ± 1·4 | IV | $C_4H_9O^-$ | Pt | H | 69 | 102 | 14·9 | |
| di-t-Butyl peroxide | $C_8H_{18}O_2$ | 7·3 | 1330 | 2·0 | III | $C_4H_9O^-$ | W | — | — | 37·0 | 20·5 | |
| Dibenzoyl peroxide | $C_{14}H_{10}O_4$ | 49·6 | 1400 | 44·0 | II | $C_7H_5O_2^-$ | Pt | — | — | — | 44·0 | |
| Water | $H_2O$ | −32·0 | 2000 | −44·0 | IV | $O^-$ | W | H | 2 × 72 | 219 | 31·0 | Note 1 |
| Hydrogen peroxide | $H_2O_2$ | 93·7 ± 4·8 | 1630 | 83·9 ± 4·8 | IV | $OH^-$ | Ir | OH | 90·6 | 47·6 | 40·9 ± 4·8 | Note 2 |
| Hydrogen peroxide | $H_2O_2$ | 49·1 ± 2·7 | 1820 | 41·8 ± 2·7 | II | $OH^-$ | Ir | — | — | — | 41·8 ± 2·7 | Note 2 |
| Nitrogen dioxide | $NO_2$ | −32·1 ± 1·7 | 1830 | −39·4 ± 1·7 | III | $O^-$ | Pt | — | — | 72 | 32·6 ± 1·7 | |
| Nitrous oxide | $N_2O$ | 38·9 | 1850 | 31·5 | II | $O^-$ | W | — | — | — | 31·5 | |
| Oxygen | $O_2$ | −20·2 | 1750 | −27·2 | III | $O^-$ | W | — | — | 118 | 31·8 | |

*Note 1*: The energy given in column $D$ is the total heat of atomization.
*Note 2*: The heat of adsorption of hydroxyl on iridium was derived from these two results.

Table 12.3  σ-Silicon Ions

| Compound | Formula | $E_T'$ | $T°K$ | $E_0'$ | Type of Split | Ion | Fil. | Res. | Q | D | $E_0$ |
|---|---|---|---|---|---|---|---|---|---|---|---|
| Silicon tetrafluoride | $F_4Si$ | 0 | 1350 | −8·0 | IV | $F_3Si^-$ | Pt | F | 51·8 ± 2·8 | 137 | 77·2 ± 2·8 |

Table 12.4  σ-Phosphorus Ions

| Compound | Formula | $E_T'$ | $T°K$ | $E_0'$ | Type of Split | Ion | Fil. | Res. | Q | D | $E_0$ | |
|---|---|---|---|---|---|---|---|---|---|---|---|---|
| Diphenylphosphine | $C_{12}H_{11}P$ | 27 | 1500 | 18·0 | IV | $C_{12}H_{10}P^-$ | Ir | H | 66 | $E - D = -48$ | — | Note 1 |
| Phosphine | $H_3P$ | 28·0 ± 5·0 | 1500 | 19·0 ± 5·0 | IV | $H_2P^-$ | W | H | 72 | 89·8 ± 5·0 | 36·8 | Note 2 |
| Diphosphine | $H_4P_2$ | 43·0 | 1500 | 36·8 | II | $H_2P^-$ | Pt | — | — | — | 36·8 | |

Note 1: If it is assumed that $E_0[(C_6H_5)_2P] = E_0[PH_2] + 1\cdot6$, analogous to $E_0[(C_6H_5)_2N] = E_0(NH_2) + 1\cdot6$ then $D[(C_6H_5)_2P{-}H] = 86\cdot4$ kcal mole⁻.
Note 2: The $D(H_2P{-}H)$ energy was derived by comparison with diphosphine; conversely, if $D[(C_6H_5)_2P{-}H]$ is assumed to be equal to $D[H_2P{-}H]$ then $E_0[(C_6H_5)_2N] = 41\cdot8 ± 5\cdot0$ kcal mole⁻.

Table 12.5  σ-Sulphur Ions

| Compound | Formula | $E_T'$ | $T$ °K | $E_0'$ | Type of Split | Ion | Fil. | Res. | $Q$ | $D$ | $E_0$ | Notes |
|---|---|---|---|---|---|---|---|---|---|---|---|---|
| Bis(trifluoromethyl) disulphide | $C_2F_6S_2$ | 41·5 ± 0·8 | 1350 | 36·1 ± 0·8 | II | $CF_3S^-$ | W | — | — | — | 36·1 ± 0·8 | |
| Ethyl mercaptan | $C_2H_6S$ | 16·4 ± 0·8 | 1450 | 7·7 ± 0·8 | IV | $C_2H_5S^-$ | Pt | H | 69 | 88·5 ± 5·0 | 27·2 ± 5·8 | |
| Dimethyl disulphide | $C_2H_6S_2$ | 1·3 ± 1·8 | 1450 | −4·5 ± 1·8 | III | $CH_3S^-$ | Pt | — | — | 69·8 | 30·4 ± 1·8 | Notes 1 and 2 |
| Dimethyl disulphide | $C_2H_6S_2$ | 12·5 ± 1·6 | <1450 | 3·8 ± 1·6 | IV | $CH_3S^-$ | Pt | MeS | 42·3 ± 4·6 | 69·8 | 31·3 ± 3·0 | Notes 1 and 2 |
| Dimethyl disulphide | $C_2H_6S_2$ | 38·5 ± 4·3 | 1600 | 32·1 ± 4·3 | II | $CH_3S^-$ | Pt | — | — | — | 32·1 ± 4·3 | Note 2 |
| Diethyl disulphide | $C_4H_{10}S_2$ | 18·3 ± 1·7 | 1440 | 9·7 ± 1·7 | IV | $C_2H_5S^-$ | Pt | $C_2H_5S$ | 42·3 ± 4·6 | 69·8 | 37·2 ± 5·3 | Notes 1 and 3 |
| Thionyl fluoride | $F_2O_2S$ | −9·4 ± 3·5 | 1700 | −16·2 ± 3·5 | III | $FO_2S^-$ | Pt | — | — | 80 | 63·8 ± 3·5 | |
| Sulphur tetrafluoride | $F_4S$ | −11·5 ± 3·4 | 1500 | −17·6 ± 3·4 | III | $F_3S^-$ | Pt | — | — | 80 | 62·4 ± 3·4 | |
| Sulphur hexafluoride | $F_6S$ | 6·6 ± 2·0 | 1430 | 0·9 ± 2·0 | III | $SF_5^-$ | Pt | — | — | 82·6 ± 2·2 | 83·5 ± 4·2 | |
| Disulphur decafluoride | $F_{10}S_2$ | 88·4 ± 0·8 | 1390 | 82·8 ± 0·8 | II | $SF_5^-$ | Pt | — | — | — | 82·8 ± 0·8 | |
| Hydrogen sulphide | $H_2S$ | 31·0 ± 2·0 | 1740 | 20·6 ± 2·0 | IV | $S^-$ | W | H | $2 \times 27$ | 165 | 41·6 ± 2·0 | |
| Hydrogen disulphide | $H_2S_2$ | 58·8 ± 0·7 | 2100 | 50·4 ± 0·7 | II | $HS^-$ | W | — | — | — | 50·4 ± 0·7 | |

Note 1: $D$(S—S) used is that quoted by Mackle[1] for dimethyl disulphide.
Note 2: The heat of adsorption of methyl sulphide on platinum is derived from these results.
Note 3: The heat of adsorption of ethyl sulphide on platinum used is that derived for methyl sulphide.

Table 12.6  Halogen Ions

| Compound | Formula | $E_T'$ | $T°K$ | $E_0'$ | Type of Split | Ion | Fil. | Res. | Q | D | $E_0$ |
|---|---|---|---|---|---|---|---|---|---|---|---|
| Hydrogen bromide | BrH | 70·0 | 1800 | 59·2 | IV | Br⁻ | W | H | 72 | 82 | 69·2 |
| Carbon tetrachloride | CCl₄ | 21·4 ± 0·3 | 1560 | 15·2 ± 0·3 | III | Cl⁻ | Pt | — | — | 69·9 ± 0·3 | 85·1 ± 0·6 |
| Fluorobenzoquinone | C₆FH₃O₂ | −26·4 ± 0·8 | 1380 | −31·9 ± 0·8 | III | F⁻ | Ir | — | — | 111·3 | 79·4 ± 0·8 |
| Hydrogen chloride +H₂ | ClH | 61·0 | 1800 | 50·2 | IV | Cl⁻ | W | H | 72 | 102 | 80·2 |
| Iodine | I₂ | 79·5 | 1650 | 72·9 | II | I⁻ | W | — | — | — | 72·9 |

**REFERENCE**

1. Mackle, H., *Tetrahedron*, **19**, 1159 (1963).

# APPENDIX I

## List of Substances Studied in the Magnetron

The use of this list, together with Appendix II, will enable the energetics of ion formation in any vapour studied to be traced back to the original experimental results. The formulae follow a systematic alphabetical listing. The mass number is used for the purpose of indexing the ions only. It is not implied that the ions have been observed in a mass spectrometer.

| Formula | Substance | Mass Number of Ions Observed |
|---|---|---|
| $AlC_{15}F_{18}H_3O_6$ | Tris(hexafluoroacetonylacetonate) aluminium | 648 |
| $BBr_3$ | Boron tribromide | 249 |
| $BF_3$ | Boron trifluoride | 66 |
| $BrC_6F_5$ | Bromopentafluorobenzene | 167, 226 |
| $BrH$ | Hydrogen bromide | 80 |
| $Br_2$ | Bromine | 80 |
| $Br_2C_6H_4$ | p-dibromobenzene | 156 |
| $Br_3CH$ | Bromoform | 173, 252 |
| $Br_4C$ | Carbon tetrabromide | 252, 332 |
| $CClF_3$ | Chlorotrifluoromethane | 69 |
| $CCl_2H_2$ | Methylene dichloride | 84 |
| $CCl_3H$ | Chloroform | 117, 118 |
| $CCl_4$ | Carbon tetrachloride | 35, 117, 152 |
| $CF_3H$ | Fluoroform | 69 |
| $CF_4$ | Carbon tetrafluoride | 69 |

(continued

| Formula | Substance | Mass Number of Ions Observed |
|---------|-----------|------------------------------|
| $CHN$ | Hydrogen cyanide | 26 |
| $CHNS$ | Thiocyanic acid | 58 |
| $CH_3I$ | Methyl iodide | 127 |
| $CH_3NO_2$ | Nitromethane | 46 |
| $CH_4O$ | Methanol | 31 |
| $CO_2$ | Carbon dioxide | 16 |
| | | |
| $C_2Cl_6$ | Hexachloroethane | 119, 234 |
| $C_2F_4$ | Tetrafluoroethylene | 50 |
| $C_2F_6$ | Hexafluoroethane | 69 |
| $C_2F_6O_2$ | Bis(trifluoromethyl)peroxide | 85 |
| $C_2F_6S_2$ | Bis(trifluoromethyl)disulphide | 101 |
| $C_2H_2$ | Acetylene | 25 |
| $C_2H_3N$ | Acetonitrile | 26 |
| $C_2H_6Hg$ | Dimethylmercury | 15 |
| $C_2H_6O$ | Ethanol | 45 |
| $C_2H_6S$ | Ethylmercaptan | 61 |
| $C_2H_6S_2$ | Dimethyl disulphide | 47 |
| $C_2N_2$ | Cyanogen | 26 |
| $C_2N_2S_2$ | Thiocyanogen | 58 |
| $C_2N_2Se_2$ | Selenocyanogen | 165 |
| $C_3F_8$ | Octafluoropropane | 169 |
| $C_3H_4$ | Allene | 39 |
| $C_3H_8O$ | *i*-propanol | 59 |
| $C_4F_8$ | Octafluorobut-2-ene | 181 |
| $C_4H_2$ | Diacetylene | 25 |
| $C_4H_2N_2$ | Fumaronitrile | 77, 78 |
| $C_4H_{10}Hg$ | Diethylmercury | 29 |
| $C_4H_{10}O$ | *n*-butanol | 73 |
| $C_4H_{10}S_2$ | Diethyl disulphide | 61 |
| $C_4H_{12}Ph$ | Tetramethyllead | 15 |
| $C_5F_6H_2O_2$ | Hexafluoroacetonylacetone | 208 |
| $C_6ClF_5$ | Chloropentafluorobenzene | 167 |
| $C_6ClH_4NO_2$ | 1-chloro-2-nitrobenzene | 122 |
| | 1-chloro-3-nitrobenzene | 122 |
| | 1-chloro-4-nitrobenzene | 122 |
| $C_6ClH_5$ | Chlorobenzene | 77 |
| $C_6Cl_2H_4$ | 1:2-dichlorobenzene | 111 |
| | 1:3-dichlorobenzene | 111 |
| | 1:4-dichlorobenzene | 111, 145 |
| $C_6Cl_3H_3$ | 1:2:3-trichlorobenzene | 145 |
| | 1:3:5-trichlorobenzene | 145 |
| $C_6Cl_4H_2$ | 1:2:3:4-tetrachlorobenzene | 179 |
| | 1:2:4:5-tetrachlorobenzene | 179 |
| $C_6Cl_4O_2$ | Chloranil | 244 |
| $C_6Cl_6$ | Hexachlorobenzene | 247 |

| Formula | Substance | Mass Number of Ions Observed |
|---|---|---|
| $C_6FH_3O_2$ | Fluorobenzoquinone | 19, 125, 126 |
| $C_6F_4O_2$ | Fluoranil | 180 |
| $C_6F_5H_2N$ | Pentafluoroaniline | 182 |
| $C_6F_6$ | Hexafluorobenzene | 167, 186 |
| $C_6H_3N_3O_6$ | 1:3:5-trinitrobenzene | 213 |
| $C_6H_4O_2$ | *p*-benzoquinone | 107, 108 |
| $C_6H_6$ | Benzene | 77 |
| $C_6H_7N$ | Aniline | 92 |
| $C_6H_{14}Hg$ | di-*n*-propylmercury | 43 |
| $C_6N_4$ | Tetracyanoethylene | 128 |
| $C_7H_8$ | Toluene | 91 |
| $C_7H_9N$ | Methylaniline | 105 |
| $C_8H_2N_2O_2$ | 2:3-dicyanobenzoquinone | 157 |
| $C_8H_4N_2$ | Phthalonitrile | 128 |
| $C_8H_6$ | Phenylacetylene | 101 |
| $C_8H_{18}Hg$ | di-*n*-butylmercury | 57 |
| $C_8H_{18}O_2$ | di-*t*-butyl peroxide | 73 |
| $C_8H_{20}Pb$ | Tetraethyllead | 15, 29 |
| $C_9HN_5$ | 2:3:5:6-tetracyanopyridine | 179 |
| $C_{10}F_4H_4$ | Tetrafluoronaphthalene | 199 |
| $C_{10}F_8$ | Octafluoronaphthalene | 253 |
| $C_{10}H_2N_4$ | 1:2:4:5-tetracyanobenzene | 178 |
| $C_{10}H_6O_2$ | Naphthoquinone | 157 |
| $C_{10}H_8$ | Azulene | 127 |
|  | Naphthalene | 127 |
| $C_{10}H_{12}O_2$ | Duroquinone | 163 |
| $C_{10}N_6$ | Hexacyanobutadiene | 204 |
| $C_{12}H_4N_4$ | 77′:88′-tetracyanoquinodimethane | 176 |
| $C_{12}H_{10}Hg$ | Diphenylmercury | 77 |
| $C_{12}H_{11}N$ | Diphenylamine | 167 |
| $C_{12}H_{11}P$ | Diphenylphosphine | 185 |
| $C_{12}H_{12}N_2$ | Diphenylhydrazine | 92 |
| $C_{12}N_6$ | Hexacyanobenzene | 228 |
| $C_{13}H_{12}$ | Diphenylmethane | 167 |
| $C_{14}Cl_2H_8O_2$ | 2:2′-dichlorobenzil | 111 |
|  | 3:3′-dichlorobenzil | 111 |
| $C_{14}H_8O_2$ | Anthraquinone | 208 |
| $C_{14}H_8N_2O_6$ | 3:3′-dinitrobenzil | 122 |
| $C_{14}H_{10}$ | Anthracene | 177 |
| $C_{14}H_{10}O_2$ | Benzil | 77 |
| $C_{14}H_{10}O_4$ | Dibenzoyl peroxide | 121 |
| $C_{14}H_{14}$ | Dibenzyl | 91 |
| $C_{14}H_{14}Hg$ | di-*p*-tolylmercury | 91 |
| $C_{14}H_{16}N_4$ | 1:4-Dimethyl-1,4-diphenyltetrazene | 105 |
| $C_{15}CrF_9H_{12}O_6$ | Tris(trifluoroacetonylacetone) chromium | 511 |

(*continued*

| Formula | Substance | Mass Number of Ions Observed |
|---|---|---|
| $C_{15}CrF_{18}H_3O_6$ | Tris(hexafluoroacetonylacetone) chromium | 673 |
| $C_{19}H_{16}$ | Triphenylmethane | 243 |
| $C_{24}H_{20}N_2$ | Tetraphenylhydrazine | 168 |
| $ClH$ | Hydrogen chloride | 35 |
| $Cl_2$ | Chlorine | 35 |
| $Cl_3HSi$ | Silicochloroform | 137 |
| $F_2O_2S$ | Thionylfluoride | 83 |
| $F_4N_2$ | Tetrafluorohydrazine | 52 |
| $F_4S$ | Sulphur tetrafluoride | 89 |
| $F_4Si$ | Silicon tetrafluoride | 85 |
| $F_6S$ | Sulphur hexafluoride | 127, 146 |
| $F_6W$ | Tungsten hexafluoride | 298 |
| $F_6U$ | Uranium hexafluoride | 352 |
| $F_{10}S_2$ | Disulphur decafluoride | 127 |
| $H_2$ | Hydrogen | 1 |
| $H_2O$ | Water | 16 |
| $H_2O_2$ | Hydrogen peroxide | 17 |
| $H_2S$ | Hydrogen sulphide | 32 |
| $H_2S_2$ | Hydrogen disulphide | 33 |
| $H_3N$ | Ammonia | 16 |
| $H_3P$ | Phosphine | 33 |
| $H_4N_2$ | Hydrazine | 16 |
| $H_4P_2$ | Diphosphine | 33 |
| $I_2$ | Iodine | 127 |
| $NO$ | Nitric oxide | 30 |
| $NO_2$ | Nitrogen dioxide | 16, 46 |
| $N_2O$ | Nitrous oxide | 16 |
| $O_2$ | Oxygen | 16 |

# APPENDIX II

## Stabilities of Negative Ions

The mass number has been calculated on the basis of the masses of the most common isotopes.

This table includes values determined by other workers using the Magnetron or other surface ionization techniques.

| Mass Number | Formula | Substrate | Affinity | | | Table | Ref. |
|---|---|---|---|---|---|---|---|
| | | | KJ | eV | kcal | | |
| 1 | H⁻ | Hydrogen | 77 | 0·80 | 18·4 | — | 1 |
| 15 | CH₃⁻ | Mercury dimethyl | 104 | 1·08 | 24·8 | 11.2 | 2 |
| | | Lead tetramethyl | 104 | 1·08 | 24·8 | 11.2 | 2 |
| | | Lead tetraethyl | 101 | 1·05 | 24·2 | 11.5 | 3 |
| 16 | O⁻ | Carbon dioxide | 133 | 1·38 | 31·8 | 12.2 | 2 |
| | | Water | 130 | 1·35 | 31·0 | 12.2 | 4 |
| | | Nitrogen dioxide | 136 | 1·42 | 32·6 | 12.2 | 5 |
| | | Nitrous oxide | 132 | 1·37 | 31·5 | 12.2 | 4 |
| | | Oxygen | 133 | 1·38 | 31·8 | 12.2 | 4 |
| | | Oxygen | 296 | 3·07 | 70·8 | — | 6 |
| | | Nitrous oxide | 225 | 2·34 | 53·8 | — | 7 |
| | NH₂⁻ | Methylamine | 116 | 1·21 | 27·8 | — | 8 |
| | | Ammonia | 108 | 1·12 | 25·7 | 12.1 | 9 |
| | | Hydrazine | 108 | 1·12 | 25·7 | 12.1 | 9 |

*(continued*

| Mass Number | Formula | Substrate | Affinity KJ | eV | kcal | Table | Ref. |
|---|---|---|---|---|---|---|---|
| 17 | OH⁻ | Hydrogen peroxide | 171 | 1·78 | 40·9 | 12.2 | 10 |
|  |  | Hydrogen peroxide | 175 | 1·82 | 41·8 | 12.2 | 10 |
| 19 | F⁻ | Fluorobenzoquinone | 332 | 3·45 | 79·4 | 12.6 | 11 |
|  |  | Potassium fluoride | 398 | 4·13 | 95·0 | — | 12 |
|  |  | Fluorine | 345 | 3·58 | 82·4 | — | 15, 16 |
|  |  | Fluorine | 344 | 3·56 | 82·1 | — | 17 |
| 25 | C₂H⁻ | Acetylene | 260 | 2·70 | 62·2 | 11.6 | 13 |
|  |  | Diacetylene | 255 | 2·65 | 61·0 | 11.3 | 14, 47 |
| 26 | CN⁻ | Hydrogen cyanide | 295 | 3·07 | 70·6 | 11.6 | 18 |
|  |  | Cyanogen | 312 | 3·24 | 74·5 | 11.3 | 18 |
|  |  | Acetonitrile | 270 | 2·80 | 64·5 | — | 8 |
|  |  | Potassium cyanide | 301 | 3·13 | 72·0 | — | 19 |
| 29 | C₂H₅⁻ | Lead tetraethyl | 87 | 0·90 | 20·7 | 11.2 | 3 |
|  |  | Mercury diethyl | 85 | 0·88 | 20·3 | 11.2 | 20 |
| 30 | NO⁻ | Nitric oxide | 79·9 | 0·83 | 19·1 | 12.1 | 21 |
| 31 | CH₃O⁻ | Methanol | 36 | 0·38 | 8·7 | 12.2 | 22 |
| 32 | S⁻ | Hydrogen sulphide | 174 | 1·81 | 41·6 | 12.5 | 23 |
| 33 | SH⁻ | Hydrogen disulphide | 211 | 2·19 | 50·4 | 12.5 | 23 |
|  | PH₂⁻ | Phosphine | 154 | 1·60 | 36·8 | 12.4 | 13 |
|  |  | Diphosphine | 154 | 1·60 | 36·8 | 12.4 | 13 |
| 35 | Cl⁻ | Hydrogen chloride | 336 | 3·48 | 80·2 | 12.6 | 24 |
|  |  | Carbon tetrachloride | 356 | 3·70 | 85·1 | 12.6 | 25 |
|  |  | Chlorine | 388 | 4·02 | 92·7 | — | 26 |
|  |  | Chlorine | 359 | 3·73 | 85·8 | — | 27 |
|  |  | Chlorine | 363 | 3·77 | 86·6 | — | 17 |
|  |  | Potassium chloride | 364 | 3·78 | 87·0 | — | 12 |
| 39 | C₃H₃⁻ | Allene | 225 | 2·34 | 53·8 | 11.7 | 13 |
| 43 | C₃H₇⁻ | Mercury di-*n*-propyl | 61 | 0·63 | 14·6 | 11.2 | 3 |
| 44 | C₂H₆N⁻ | Dimethylamine | 100 | 1·04 | 24·0 | 12.1 | 13 |
|  |  | Tetramethyltetrazene | 100 | 1·04 | 24·0 | 12.1 | 13 |
| 45 | C₂H₅O⁻ | Ethanol | 57 | 0·59 | 13·7 | 12.2 | 22 |
| 46 | NO₂⁻ | Nitrogen dioxide | 377 | 3·91 | 90·0 | 12.1 | 5 |
|  |  | Nitromethane | 385 | 4·00 | 92·0 | — | 8 |
|  | NO₂⁻* | Nitrogen dioxide | 000 | 0·00 | 0·0 | — | 5 |
| 47 | CH₃S⁻ | Dimethyl disulphide | 127 | 1·32 | 30·4 | 12.5 | 28 |
|  |  | Dimethyl disulphide | 131 | 1·36 | 31·3 | 12.5 | 28 |
|  |  | Dimethyl disulphide | 134 | 1·39 | 32·1 | 12.5 | 28 |

| Mass Number | Formula | Substrate | Affinity KJ | eV | kcal | Table | Ref. |
|---|---|---|---|---|---|---|---|
| 50 | CF₂⁻ | Tetrafluoroethylene | 255 | 2·65 | 61·0 | 11.4 | 13 |
| 52 | NF₂⁻ | Tetrafluorohydrazine | 289 | 3·00 | 69·0 | 12.1 | 29 |
| 57 | C₄H₉⁻ | Mercury di-*n*-butyl | 57 | 0·59 | 13·6 | 11.2 | 3 |
| 58 | SCN⁻ | Thiocyanic acid | 209 | 2·17 | 49·9 | 12.1 | 18 |
|  |  | Thiocyanogen | 209 | 2·17 | 49·9 | 12.1 | 18 |
| 59 | C₃H₇O⁻ | *i*-propanol | 65 | 0·67 | 15·5 | 12.2 | 22 |
| 61 | C₂H₅S⁻ | Ethyl mercaptan | 114 | 1·18 | 27·2 | 12.5 | 28 |
|  |  | Diethyldisulphide | 156 | 1·62 | 37·2 | 12.5 | 28 |
| 66 | BF₃⁻ | Boron trifluoride | 255 | 2·65 | 61·0 | 7.1 | 29 |
| 69 | CF₃⁻ | Hexafluoroethane | 194 | 2·01 | 46·3 | 11.5 | 13 |
|  |  | Fluoroform | 179 | 1·85 | 42·7 | 11.8 | 13 |
|  |  | Fluoroform | 213 | 2·21 | 51·0 | 11.5 | 13 |
|  |  | Fluoroform | 176 | 1·82 | 42·0 | 11.2 | 13 |
|  |  | Carbon tetrafluoride | 179 | 1·85 | 42·7 | 11.8 | 13 |
|  |  | Bromotrifluoromethane | 171 | 1·78 | 41·0 | 11.8 | 13 |
|  |  | Chlorotrifluoromethane | 167 | 1·74 | 40·0 | 11.8 | 13 |
| 73 | C₄H₉O⁻ | *n*-Butanol | 62 | 0·65 | 14·9 | 12.2 | 22 |
|  |  | di-*t*-Butyl peroxide | 86 | 0·89 | 20·6 | 12.2 | 30 |
| 77 | C₄HN₂⁻ | Fumaronitrile | 168 | 1·74 | 40·1 | 11.7 | 31 |
|  | C₆H₅⁻ | Chlorobenzene | 207 | 2·15 | 49·4 | 11.7 | 32 |
|  |  | Benzene | 237 | 2·46 | 56·7 | 11.7 | 33 |
|  |  | Benzene | 231 | 2·39 | 55·1 | 11.7 | 34 |
|  |  | Benzil | 207 | 2·15 | 49·4 | 11.1 | 34 |
|  |  | Mercury diphenyl | 218 | 2·27 | 52·2 | 11.7 | 34 |
| 78 | C₄H₂N₂⁻ | Fumaronitrile | 72 | 0·75 | 17·2 | 8.1 | 31 |
| 79 | Br⁻ | Potassium bromide | 351 | 3·65 | 84·0 | — | 12 |
|  |  | Hydrogen bromide | 289 | 3·00 | 69·2 | 12.6 | 4 |
|  |  | Bromine | 335 | 3·47 | 80·0 | — | 35 |
|  |  | Bromine | 368 | 3·82 | 88·0 | — | 36 |
|  |  | Bromine | 339 | 3·52 | 81·0 | — | 37 |
|  |  | Bromine | 339 | 3·52 | 80·9 | — | 17 |
| 83 | SO₂F⁻ | Thionyl fluoride | 267 | 2·77 | 63·8 | 12.5 | 38 |
| 84 | CCl₂H₂⁻ | Methylene dichloride | 126 | 1·31 | 30·1 | 7.1 | 25 |

*(continued)*

| Mass Number | Formula | Substrate | Affinity | | | Table | Ref. |
|---|---|---|---|---|---|---|---|
| | | | KJ | eV | kcal | | |
| 85 | $F_3Si^-$ | Silicon tetrafluoride | 323 | 3·35 | 77·2 | 12.3 | 13 |
| | $CF_3O^-$ | Bis(trifluoromethyl) peroxide | 131 | 1·35 | 31·2 | 12.2 | 13 |
| 89 | $F_3S^-$ | Sulphur tetrafluoride | 261 | 2·71 | 62·4 | 12.5 | 38 |
| 91 | $C_7H_7^-$ | Toluene | 79 | 0·82 | 18·9 | 11.8 | 34 |
| | | Dibenzyl | 80 | 0·84 | 19·2 | 11.2 | 34 |
| | | Dibenzyl | 73 | 0·76 | 17·4 | 11.2 | 39 |
| | | Mercury di-*p*-tolyl | 293 | 2·52 | 58·1 | 11.7 | 34 |
| 92 | $C_6H_6N^-$ | Diphenylhydrazine | 149 | 1·55 | 35·7 | 12.1 | 13 |
| | | Aniline | 149 | 1·55 | 35·7 | 12.1 | 13 |
| 101 | $C_8H_5^-$ | Phenylacetylene | 277 | 2·88 | 66·3 | 11.6 | 13 |
| | $CF_3S^-$ | Bis(trifluoromethyl) disulphide | 174 | 1·80 | 41·5 | 12.5 | 28 |
| 105 | $C_7H_8N^-$ | 1:4-Dimethyl-1:4-diphenyltetrazene | 126 | 1·30 | 30·0 | 12.1 | 13 |
| | | Methylaniline | 126 | 1·30 | 30·0 | 12.1 | 13 |
| | $SeCN^-$ | Selenocyanogen | 255 | 2·64 | 60·9 | 12.1 | 30 |
| 107 | $C_6H_3O_2^-$ | *p*-Benzoquinone | 192 | 2·00 | 46·0 | 11.7 | 11 |
| 108 | $C_6H_4O_2^-$ | *p*-Benzoquinone | 129 | 1·34 | 30·9 | 8.1 | 11 |
| 111 | $C_6ClH_4^-$ | 1:2-Dichlorobenzene | 218 | 2·27 | 52·3 | 11.7 | 32 |
| | | 1:2-Dichlorobenzene | 223 | 2·32 | 53·4 | 11.7 | 32 |
| | | 2:2-Dichlorobenzil | 234 | 2·43 | 55·9 | 11.1 | 32 |
| | | 1:3-Dichlorobenzene | 200 | 2·08 | 47·9 | 11.7 | 32 |
| | | 1:3-Dichlorobenzene | 196 | 2·04 | 46·9 | 11.7 | 32 |
| | | 3:3-Dichlorobenzil | 223 | 2·31 | 53·3 | 11.1 | 32 |
| | | 1:4-Dichlorobenzene | 223 | 2·32 | 53·4 | 11.7 | 32 |
| | | 1:4-Dichlorobenzene | 226 | 2·35 | 54·1 | 11.7 | 32 |
| 117 | $CCl_3^-$ | Chloroform | 117 | 1·22 | 28·0 | 11.8 | 25 |
| | | Carbon tetrachloride | 117 | 1·22 | 28·0 | 11.8 | 25 |
| 118 | $CCl_3H^-$ | Chloroform | 164 | 1·70 | 39·2 | 7.1 | 25 |
| 121 | $C_7H_5O_2^-$ | Dibenzoylperoxide | 184 | 1·91 | 44·0 | 12.2 | 13 |
| 122 | $C_6H_4NO_2^-$ | 1-Chloro-2-nitrobenzene | 226 | 2·35 | 54·1 | 11.7 | 32 |
| | | 1-Chloro-3-nitrobenzene | 155 | 1·61 | 37·1 | 11.7 | 32 |
| | | 1-Chloro-3-nitrobenzene | 161 | 1·68 | 38·6 | 11.7 | 32 |
| | | 3:3'-Dinitrobenzil | 164 | 1·70 | 39·2 | 11.1 | 32 |
| | | 1-Chloro-4-nitrobenzene | 235 | 2·44 | 56·1 | 11.7 | 32 |

| Mass Number | Formula | Substrate | Affinity KJ | eV | kcal | Table | Ref. |
|---|---|---|---|---|---|---|---|
| 125 | $C_6FH_2O_2^-$ | Fluorobenzoquinone | 231 | 2·39 | 55·1 | 11.7 | 11 |
| 126 | $C_6FH_3O_2^-$ | Fluorobenzoquinone | 208 | 2·16 | 49·8 | 8.1 | 11 |
| 127 | $C_{10}H_7^-$ | Azulene | 256 | 2·66 | 61·3 | 11.4 | 33 |
|  |  | Naphthalene | 213 | 2·21 | 51·0 | 11.7 | 21 |
|  | $SF_5^-$ | Sulphur hexafluoride | 349 | 3·63 | 83·5 | 12.5 | 40 |
|  |  | Disulphur decafluoride | 346 | 3·60 | 82·8 | 12.5 | 40 |
|  | $I^-$ | Potassium iodide | 318 | 3·30 | 76·0 | — | 12 |
|  |  | Iodine | 305 | 3·17 | 72·9 | 12.6 | 4 |
|  |  | Iodine | 316 | 3·28 | 75·6 | — | 41 |
|  |  | Iodine | 303 | 3·14 | 72·4 | — | 42 |
|  |  | Iodomethane | 312 | 3·24 | 74·5 | — | 8 |
|  |  | Iodine | 314 | 3·26 | 75·0 | — | 43 |
|  |  | Iodine | 307 | 3·18 | 73·2 | — | 17 |
| 128 | $C_6N_4^-$ | Tetracyanoethylene | 278 | 2·88 | 66·4 | 8.1 | 39 |
|  | $C_8H_4N_2^-$ | *o*-Phthalonitrile | 106 | 1·04 | 23·9 | 8.1 | 31 |
| 145 | $C_6Cl_2H_3^-$ | 1:4-Dichlorobenzene | 197 | 2·05 | 47·2 | 11.7 | 39 |
|  |  | 1:2:3-Trichlorobenzene | 258 | 2·68 | 61·6 | 11.7 | 32 |
|  |  | 1:3:5-Trichlorobenzene | 216 | 2·24 | 51·6 | 11.7 | 32 |
| 146 | $SF_6^-$ | Sulphur hexafluoride | 138 | 1·43 | 33·0 | 7.1 | 40 |
| 152 | $CCl_4^-$ | Carbon tetrachloride | 198 | 2·06 | 47·4 | 7.1 | 25 |
| 155 | $BrC_6H_4^-$ | *p*-Dibromobenzene | 251 | 2·61 | 60·0 | 11.4 | 39 |
|  |  | *p*-Dibromobenzene | 248 | 2·58 | 59·3 | 11.7 | 44 |
| 157 | $C_8HN_2O_2^-$ | 2:3-Dicyanobenzo-quinone | 175 | 1·82 | 41·9 | 11.7 | 11 |
|  | $C_{10}H_5O_2^-$ | 1:4-Naphthoquinone | 212 | 2·20 | 50·7 | 11.7 | 33 |
| 163 | $C_{10}H_{11}O_2^-$ | Duroquinone | 75 | 0·78 | 18·0 | 11.8 | 11 |
|  |  | Duroquinone | 77 | 0·80 | 18·5 | 11.8 | 11 |
| 167 | $C_6F_5^-$ | Bromopentafluoro-benzene | 267 | 2·78 | 63·9 | 11.7 | 33 |
|  |  | Chloropentafluoro-benzene | 260 | 2·70 | 62·2 | 11.7 | 33 |
|  |  | Hexafluorobenzene | 265 | 2·75 | 63·3 | 11.7 | 33 |
|  | $C_{13}H_{11}^-$ | Diphenylmethane | 82 | 0·85 | 19·6 | — | 2 |
| 168 | $C_{12}H_{10}N^-$ | Diphenylamine | 114 | 1·19 | 27·3 | 12.1 | 13 |
|  |  | Tetraphenylhydrazine | 114 | 1·19 | 27·3 | 12.1 | 13 |

(*continued*)

| Mass Number | Formula | Substrate | Affinity KJ | eV | kcal | Table | Ref. |
|---|---|---|---|---|---|---|---|
| 169 | $C_3F_7{}^-$ | Octafluoropropane | 192 | 1·99 | 45·9 | 11.8 | 38 |
| 171 | $Br_2CH^-$ | Bromoform | 187 | 1·94 | 44·7 | 11.5 | 45 |
| 176 | $C_{12}H_4N_4{}^-$ | 77′:88′-Tetracyano-quinodimethane | 272 | 2·83 | 65·1 | 8.1 | 31 |
| 177 | $C_{14}H_9{}^-$ | Anthracene | 203 | 2·11 | 48·6 | 11.7 | 33 |
| 178 | $C_{10}H_2N_4{}^-$ | 1:2:4:5-Tetracyano-benzene | 207 | 2·15 | 49·5 | 8.1 | 31 |
| 179 | $C_9HN_5{}^-$ | 2:3:5:6-Tetracyano-pyridine | 204 | 2·12 | 48·7 | 8.1 | 31 |
|  | $C_6Cl_3H_2{}^-$ | 1:2:3:4-Tetrachloro-benzene | 234 | 2·43 | 55·9 | 11.7 | 32 |
|  |  | 1:2:4:5-Tetrachloro-benzene | 246 | 2·56 | 58·9 | 11.7 | 32 |
| 180 | $C_6F_4O_2{}^-$ | Fluoranil | 218 | 2·27 | 52·2 | 8.1 | 33 |
| 181 | $C_4F_7{}^-$ | Octafluorobut-2-ene | 259 | 2·69 | 61·9 | 11.7 | 38 |
| 182 | $C_6F_5NH^-$ | Pentafluoroaniline | 159 | 1·66 | 38·1 | 12.1 | 13 |
| 185 | $C_{12}H_{10}P^-$ | Diphenyl phosphine | 175 | 1·81 | 41·8 | 12.4 | 13 |
| 186 | $C_6F_6{}^-$ | Hexafluorobenzene | 115 | 1·20 | 27·6 | 8.1 | 33 |
| 199 | $C_{10}F_4H_3{}^-$ | Tetrafluoronaphthalene | 208 | 2·16 | 49·8 | 11.7 | 33 |
|  | $C_2Cl_5{}^-$ | Hexachloroethane | 150 | 1·55 | 35·8 | 11.8 | 25 |
| 204 | $C_{10}H_6{}^-$ | Hexacyanobutadiene | 312 | 3·24 | 74·7 | 8.1 | 31 |
| 208 | $C_5F_6H_2O_2{}^-$ | Hexafluoroacetylacetone | 335 | 3·47 | 80·0 | — | 13 |
|  | $C_{14}H_8O_2{}^-$ | Anthraquinone | 111 | 1·15 | 26·5 | 8.1 | 33 |
| 213 | $C_6H_3N_3O_6{}^-$ | 1:3:5-Trinitrobenzene | 253 | 2·63 | 60·5 | 8.1 | 39 |
| 225 | $C_6F_4Br^-$ | Bromopentafluoro-benzene | 265 | 2·75 | 63·3 | 11.7 | 33 |
| 228 | $C_{12}N_6{}^-$ | Hexacyanobenzene | 239 | 2·48 | 57·2 | 8.1 | 31 |
| 237 | $C_2Cl_6{}^-$ | Hexachloroethane | 137 | 1·42 | 32·8 | 7.1 | 25 |
| 243 | $C_{19}H_{15}{}^-$ | Triphenylmethane | 78 | 0·80 | 18·1 | — | 2 |
| 246 | $C_6Cl_4O_2{}^-$ | Chloranil | 231 | 2·40 | 55·3 | 8.1 | 11 |
| 247 | $C_6Cl_5{}^-$ | Hexachlorobenzene | 266 | 2·76 | 63·6 | 11.7 | 33 |
| 249 | $Br_3C^-$ | Bromoform | 173 | 1·80 | 41·4 | 11.8 | 45 |
|  |  | Carbon tetrabromide | 148 | 1·53 | 35·3 | 11.8 | 45 |
|  |  | Carbon tetrabromide | 181 | 1·87 | 43·2 | 11.5 | 45 |

| Mass Number | Formula | Substrate | Affinity | | | Table | Ref. |
|---|---|---|---|---|---|---|---|
| | | | KJ | eV | kcal | | |
| 253 | $C_{10}F_7^-$ | Octafluoronaphthalene | 244 | 2·54 | 58·4 | 11.7 | 33 |
| 298 | $WF_6^-$ | Tungsten hexafluoride | 264 | 2·74 | 63·0 | — | 13 |
| 328 | $CBr_4^-$ | Carbon tetrabromide | 195 | 2·06 | 46·7 | 7.1 | 45 |
| 353 | $UF_6^-$ | Uranium hexafluoride | 280 | 2·91 | 67·0 | — | 46 |
| 511 | $C_{15}H_{12}F_9 \cdot O_6Cr^-$ | $Cr^{III}$trifluoroacetyl-acetonate | 192 | 2·00 | 46·0 | — | 13 |
| 648 | $C_{15}H_3F_{18} \cdot O_6Al^-$ | $Al^{III}$hexafluoroacetyl-acetonate | 264 | 2·74 | 63·0 | — | 13 |
| 673 | $C_{10}H_3F_{18} \cdot O_6Cr^-$ | $Cr^{III}$hexafluoroacetyl-acetonate | 314 | 3·26 | 75·0 | — | 13 |

## REFERENCES

1. Dukelskii, V. M. and Khvostenko, V. I., *J. Exptl. Theoret. Phys.*, *USSR*, **37**, 657 (1959).
2. Gaines, A. F. and Page, F. M., *Unpublished work*.
3. Page, F. M., *Proc. 8th Symp. Combustion*, **57**, Williams and Wilkins, Baltimore, 1962.
4. Page, F. M., *Trans. Faraday Soc.*, **57**, 359 (1961).
5. Farragher, A. L., Page, F. M. and Wheeler, R. C., *Disc. Faraday Soc.*, **37**, 203 (1964).
6. Mayer, J. E. and Vier, D. T., *J. Chem. Phys.*, **12**, 28 (1944).
7. Metlay, M. and Kimball, G. E., *J. Chem. Phys.*, **16**, 774 (1948).
8. Ritchie, B. and Wheeler, R. C., *J. Phys. Chem.*, **70**, 113 (1966).
9. Page, F. M., *Trans. Faraday Soc.*, **57**, 1254 (1961).
10. Kay, J. and Page, F. M., *Trans. Faraday Soc.*, **62**, 3081 (1966).
11. Farragher, A. L. and Page, F. M., *Trans. Faraday Soc.*, **62**, 3072 (1966).
12. Dukelskii, V. M. and Ionov, N. I., *J. Exptl. Theoret. Phys.*, *USSR*, **10**, 1248 (1940).
13. Goode, G. C., *Unpublished work*.
14. Buttfield, E., *B.Sc. Thesis*, The University of Aston in Birmingham, 1968.
15. Metlay, M. and Kimball, G. E., *J. Chem. Phys.*, **16**, 779 (1948).
16. Bernstein, R. B. and Metlay, M., *J. Chem. Phys.*, **19**, 1612 (1951).
17. Bailey, T. L., *J. Chem. Phys.*, **28**, 792 (1958).
18. Napper, R. and Page, F. M., *Trans. Faraday Soc.*, **59**, 1086 (1963).
19. Bakulina, I. N. and Ionov, N. I., *Doklady Akad. Nauk. SSSR*, **99**, 1023 (1954).
20. Page, F. M., *Unpublished work*.
21. Rees, C. W. L., *Dip. Tech. Thesis*, College of Advanced Technology, Birmingham, 1964.
22. Baldwin, C. J., *Dip. Tech. Thesis*, College of Advanced Technology, Birmingham, 1964.
23. Ansdell, D. A. and Page, F. M., *Trans. Faraday Soc.*, **58**, 1084 (1962).

24. Page, F. M., *Trans. Faraday Soc.*, **56**, 1742 (1960).
25. Gaines, A. F., Kay, J. and Page, F. M., *Trans. Faraday Soc.*, **62**, 874 (1966).
26. Mayer, J. E. and Mitchell, J. J., *J. Chem. Phys.*, **8**, 282 (1940).
27. Mayer, J. E. and McCallum, K. J., *J. Chem. Phys.*, **11**, 56 (1943).
28. Cowley, L. T., *Dip. Tech. Thesis*, College of Advanced Technology, Birmingham, 1965.
29. Kay, J., *U.S. Army Final Tech. Rept.* (January, 1965).
30. Newman, R. N., *B.Sc. Thesis*, The University of Aston in Birmingham, 1967.
31. Farragher, A. L. and Page, F. M., *Trans. Faraday Soc.*, **63**, 2369 (1967).
32. Moss, A. D., *M.Sc. Thesis*, University of Wales, 1967.
33. Burdett, M., *Ph.D. Thesis*, University of Birmingham, 1968.
34. Gaines, A. F., and Page, F. M., *Trans. Faraday Soc.*, **59**, 1266 (1963).
35. Weissblatt, H. B., *Dissertation*, Johns Hopkins University, 1938.
36. Glockler, G. and Calvin, M., *J. Chem. Phys.*, **4**, 492 (1936).
37. Doty, P. M. and Mayer, J. E., *J. Chem. Phys.*, **12**, 323 (1944).
38. Spong, P. L., *Unpublished work*.
39. Farragher, A. L., *Ph.D. Thesis*, The University of Aston in Birmingham, 1966.
40. Kay, J. and Page, F. M., *Trans. Faraday Soc.*, **60**, 1042 (1964).
41. Mayer, J. E. and Sutton, P. P., *J. Chem. Phys.*, **2**, 145 (1934).
42. Mayer, J. E. and Sutton, P. P., *J. Chem. Phys.*, **3**, 20 (1935).
43. Glockler, G. and Calvin, M., *J. Chem. Phys.*, **3**, 771 (1935).
44. Perrett, R., *Unpublished work*.
45. Kay, J., *Unpublished work*.
46. Walters, D., *B.Sc. Thesis*, The University of Aston in Birmingham, 1968.
47. Powell, D. C., *Unpublished work*.

# APPENDIX III

## Compilations of Data

Pritchard, H. O., *Chem. Rev.*, **52**, 529 (1953).

Field, F. H. and Franklin, J. L., *Electron Impact Phenomena*, Academic Press, New York, 1957.

Buchel'nikova, N. S., *Usp. fiz. nauk*, **65**, 351 (1958); Translated by A. L. Monks, A.E.C. Tr. 3657, Oak Ridge (1959).

Vedeneyev, V. I., Gurvich, L. V., Kondrat'yev, V. N., Medvedev, V. A. and Frankevich, Ye. L., *Energiya razryva khimicheskikh suyazei*, Academy of Sciences, U.S.S.R., Moscow, 1962; Translated by Scripta Technica Ltd., as *Bond Energies, Ionisation Potentials and Electron Affinities*, Arnold, London, 1966.

Schexnayder, Charles J. Jr., *Tabulated Values of Bond Dissociation Energies, Ionization Potentials and Electron Affinities*, N.A.S.A. Technical Note D 1791, Washington, 1963.

# Author Index

The numbers in square brackets refer to the reference number where the author's work is quoted in full at the end of the chapter

147

# Subject Index